SOIL EROSION

SOIL EROSION: PROCESSES, PREDICTION, MEASUREMENT, AND CONTROL

Dr. Terrence J. Toy
Department of Geography
University of Denver
Denver, Colorado

Dr. George R. Foster
Hydraulic Research Engineer
Bryan, Texas

Dr. Kenneth G. Renard
Hydraulic Research Engineer
Tucson, Arizona

John Wiley & Sons, Inc.

Soil Erosion : Processes, Prediction, Measurement, and Control

First Indian Reprint: 2013

Authorized reprint by Wiley India Pvt. Ltd., 4435-36/7, Ansari Road, Daryaganj, New Delhi – 110002.

Published by John Wiley & Sons Inc., New York.

Printed at: *Syndicate Binders, NOIDA*

ISBN: 978-81-265-3695-5

To

Linda
(TJT)

Sherry
(GRF)

Virginia
(KGR)

For all of their patience and understanding
in the past, present, and future

Contents

Preface

This textbook is about soil-erosion processes. It is intended for students in universities or colleges and in the field working as conservationists or reclamationists. Students of erosion may come from a variety of academic disciplines with a variety of classroom and field experiences. Although this book is aimed toward upper-level undergraduate students, beginning graduate students, and others with equivalent education, only a sincere desire to understand soil erosion and erosion processes is assumed.

The book is structured into five parts. In the first part (Chapters 1 and 2) we describe the importance of erosion as a problem of global dimensions with local solutions and place erosion within the context of the environmental system operating at the local scale. In the second part (Chapters 3 and 4) we discuss types of erosion, processes within those types, and the relationships among erosion processes and environmental characteristics. In the third part (Chapters 5 and 6) we use the understanding of erosion processes to explain how erosion-prediction technologies (erosion models) are developed and how erosion research is conducted in the field and in the laboratory. An understanding of erosion processes, erosion-prediction technologies, and erosion research provide a foundation for an understanding of erosion-control concepts and practices (Chapter7). Land conservation (Chapter 8) is the management of erosion processes within a cultural context. Opportunities for conservation vary worldwide. Conservation planning, planning tools, and examples of conservation programs for selected land uses are presented in this chapter. In the final chapter (Chapter 9) we summarize the essential lessons from the preceding chapters and present our perspective concerning the future of erosion problems and erosion control. Collectively, the authors have over 100 years of experience in erosion research, conservation, and reclamation. We are unashamed to offer our viewpoints. In addition, the book provides appendices on soils and hydrology for those needing additional background.

Finally, another appendix lists numerous Web sites where additional erosion information can be found.

Land uses that result in soil exposure to the forces of water and wind cause acceleration of erosion rates. In this book we stress erosion and erosion-control principles independent of land use. With a knowledge of the principles, the conservationist and reclamationist can understand what will happen to the land as the consequence of disturbance and what needs to be done to control erosion at a particular site. This is intended to be a textbook, not a reference book, although many important works are cited in the list of references following the appendices.

Soil erosion is a very serious environmental problem in many parts of the world. Although the fundamentals of erosion control are well known, cultural and socioeconomic factors often preclude the application of erosion-control practices on the ground. We ask that instructors using this book convey an appropriate sense of urgency concerning erosion control and convey to students the excitement and satisfaction derived from erosion research and mitigation of erosion problems. We ask that the students of erosion recognize that science is not a vocation but a way of life and that erosion science is an opportunity to make the world a better place.

TJT
GRF
KGR

Acknowledgments

Several of our colleagues provided valuable assistance in the preparation of this textbook, especially Richard Hadley, Doug Helm, C. Huang, Keith McGregor, Don Meyer, Karl Musser, Mary Nichols, Waite Osterkamp, Matt Römkens, Roger Simanton, Ed Skidmore, Glenn Weesies, Daniel Yoder, and Ted Zobeck. The Department of Geography at the University of Denver, David Longbrake, Chair, supported this project in various ways during the past two years. We want them all to know that we sincerely appreciate their help.

In addition, we would like to recognize the contributions to soil erosion science of Walt Wischmeier. He had the ability, not only to lead and inspire soil erosion research, but also to transform that information into forms that proved useful all over the world. He will be long remembered.

1

Introduction

Imagine a small group of processes that have operated on the land since the first rains and winds millions of years ago and in all but the coldest and driest regions. These processes are largely responsible for the shape of the Earth's land surface. They divide the land into drainage basins, sculpt the mountains and the valleys, and form the hillslopes and stream channels. They are capable of stripping the fertile topsoil from the land, topsoil that was tens, hundreds, or even thousands of years in the making. They are capable of destroying the productivity of the land in just a few years or even months, quite literally taking the food from the mouths of men, women, children, and other fauna. This small group of processes is known as *erosion and sedimentation*, and includes detachment, entrainment, transportation, and deposition of soil and other earth materials.

On the basis of its temporal and spatial ubiquity, erosion qualifies as a major, quite possibly *the* major, environmental problem worldwide. Due to its temporal and spatial ubiquity, together with its numerous impacts, erosion is an essential research topic for physical and social scientists alike. Serious efforts by farmers, miners, contractors, and the personnel of several government agencies are necessary to protect the soil by means of effective erosion-control programs and to minimize both on- and off-site damage resulting from erosion and sedimentation.

Today, the rate of soil erosion exceeds the rate of soil formation over wide areas resulting in the depletion of soil resources and productive potential (Figure 1.1). This disparity between erosion and soil-formation rates usually is the result of human activities. As the global population increases and the demands for food, shelter, and standard-of-living expectations increase, soil depletion proceeds at faster rates and over wider areas.

Figure 1.1 Accelerated erosion on agricultural lands in south-central Iowa. (Courtesy of USDA, NRCS.)

In the United States alone, about 57.3 million acres (23.2 million hectares) of fragile highly erodible cropland was determined to experience excessive erosion, and about 50.5 million acres (20.4 million hectares) of non–highly erodible cropland was determined to have erosion that exceeded the tolerable soil-loss rate [U.S. Department of Agriculture (USDA), Natural Resources Conservation Service (NRCS), 1997b revised, 2000]. The soil-loss tolerance is the "maximum level of soil erosion that will permit a high level of crop productivity to be sustained economically and indefinitely" (Wischmeier and Smith, 1978); the utility and limitations of this concept are discussed in Chapter 8. The tragedy, of course, is that the technology exists to control erosion rates in nearly all circumstances.

During the past 30 years, many studies have documented the magnitude of soil-erosion problems, expressed as *billions* of tons of eroded soil or *billions* of dollars of erosion and sedimentation damage each year (summaries by Lal, 1994a; Morgan, 1991). Most authors acknowledge that these data are imprecise, due to the temporal and spatial variability of erosion processes, the paucity of accurate erosion measurements, extrapolation of data from small plots to continental scales, and the conversion of erosion and sedimentation rates into monetary units (Boardman, 1998; Crosson, 1995; Lal, 1994a; Osterkamp et al., 1998; Pimental et al., 1995;

Ribaudo, 1986). Nevertheless, by any measure, soil erosion is a monumental problem throughout the world, threatening ecosystems and human well-being.

The average erosion rates for very large areas misrepresents the true dimensions of erosion problems. An average value disguises the areas of low and high erosion rates. Some parts of a large area frequently experience low erosion rates that may not be problematic or may be controlled through minor and inexpensive modifications of cultural practices. Other parts of a large area often experience high erosion rates that require substantial efforts and resources to control erosion. Often, a small proportion of a land area is responsible for a large proportion of the total erosion and sediment yield. Erosion control targeted toward the areas with the highest rates can markedly reduce erosion averages.

> Soil erosion is an issue where the adage "think globally, act locally," is clearly *apropos*. Think globally, because soil erosion is a common problem that has, does, and will continue to impact the global community. Act locally, because effective erosion control requires action at the hillslope, field, stream channel, and upland watershed scales.

In most cases, long-term soil productivity and long-term sustainable agriculture require soil-erosion rates that do not exceed soil-formation rates. Soil productivity is the capacity of a soil, in its normal environment, to produce a particular plant or sequence of plants under a specified management system (National Soil Erosion–Soil Productivity Research Planning Committee, 1981). Soil productivity can be maintained and even enhanced, at least in the short term, through the use of high-yield plant varieties, pesticides, and fertilizers. These practices, however, are not economically feasible in some cases, reducing farm-business profitability, and sometimes cause other environmental problems, such as water pollution due to the transport of pesticides and fertilizers in the runoff from fields. Further, the transfer of these technologies to the "developing world" often is limited by economic and other cultural conditions.

The development and management of effective erosion-control programs require a thorough understanding of erosion processes, the ability to measure and estimate erosion rates accurately, and a knowledge of the theory and practice of erosion-control techniques. The goal of this introductory soil-erosion textbook is to (1) provide a fundamental knowledge of erosion processes, measurement, estimation, and control; (2) identify sources of more detailed information; and (3) lay the foundation for a career in erosion research and soil conservation.

PHYSICAL AND ECONOMIC SIGNIFICANCE OF EROSION

Soil erosion affects the land and its inhabitants in various direct and indirect ways. In this section, the physical and economic ramifications of erosion are discussed and social issues are addressed later in the chapter.

Changes in Soil Properties

Soil properties are the product of pedogenic (soil-forming) processes, frequently modified by human activities. Properties such as material strength, infiltration capacity, and plant productivity are altered by erosion processes (Appendix A). Soils possess strength properties that largely determine the ability to resist stresses. These properties often change in the long term as the result of weathering, pedogenic processes, and decomposition of organic matter, as well as in the short term as a result of seasonal climate conditions. Accelerated erosion removes the upper layer (A-horizon) of the soil, exposing the underlying layer (B-horizon) that may possess different strength properties and, hence, different abilities to resist the stresses imposed by gravitational forces, raindrop impact, and surface runoff.

Another important property of soils is the *infiltration capacity*, defined as the maximum rate at which water enters the soil. The infiltration capacity of the soil divides precipitation into soil moisture, groundwater, and surface runoff (Appendix B). Soil moisture binds soil particles together, reducing wind erosion rates. Water flowing beneath the surface produces pore-water pressures that reduce the friction between soil particles, making the particles more susceptible to gravitational and erosive forces. Subsurface flows may emerge downslope as seepage and contribute to surface flows. Once the precipitation rate exceeds the infiltration capacity of the soil, runoff collects and flows across the land surface, generating the hydraulic forces that erode and transport sediment from hillslopes and through stream channels. The infiltration capacity of the A-horizon often is substantially higher than the infiltration capacity of the B-horizon. When erosion removes the A-horizon, exposing the

B-horizon, a precipitation event of given intensity produces greater runoff volume and velocity, due to the lower infiltration capacity, and causes higher erosion rates, depending on the susceptibility (erodibility) of the B-horizon material to erosive forces.

Soils possess physical and chemical properties that strongly influence vegetation growth. Productivity suffers from too much or too little water. Slow-draining soils may become waterlogged, akin to overwatered household plants, making the soil suitable only for hydrophytic (wetland) plant types. Fast-draining soils may not retain adequate water to support other than xerophytic (desert land) plant types. The water-holding capacity of a soil is related to the particle-size composition. Fine-textured soils possess greater surface area to which water molecules can adsorb and be stored than do coarse-textured soils. As a rule of thumb, soils composed of 70% or more sand-size particles are considered to be droughty. When fine-size particles are removed from the soil by water and wind erosion processes, the water-holding capacity of the soil decreases. The decrease in water-holding capacity adversely affects plant growth if water becomes a limiting factor. In addition, the reduction of soil depth due to erosion decreases the volume of soil involved in water and nutrient storage.

Fertility is the capacity of a soil to provide the quantities and balances of elements and compounds necessary for plant development. Plant nutrients are stored and cycled through the upper layers of soil. Removal of these layers by erosion diminishes the quantities of nutrients available for plant use. Some plant nutrients and pesticides (including insecticides, herbicides, fungicides), whether naturally occurring or applied by the farmer or reclamation specialist, are adsorbed to sediment particles and transported from the site by runoff and erosion. Figure 1.2 shows the spatial variability of potential pesticide runoff in the United States.

Economic Consequences of Erosion

The economic consequences of erosion can be examined at local (field, farm, or construction site), regional, and national scales. Changes in the physical and chemical properties of the soil affect farm-business profitability by reducing crop yields or increasing management requirements to maintain yields. According to the USDA, NRCS (1997a), soil erosion continues to threaten the productive capacity of nearly one-third of the cropland and at least one-fifth of all rangeland in the United States. Soil erosion reduces crop yields by reducing soil organic matter, water-holding capacity, rooting

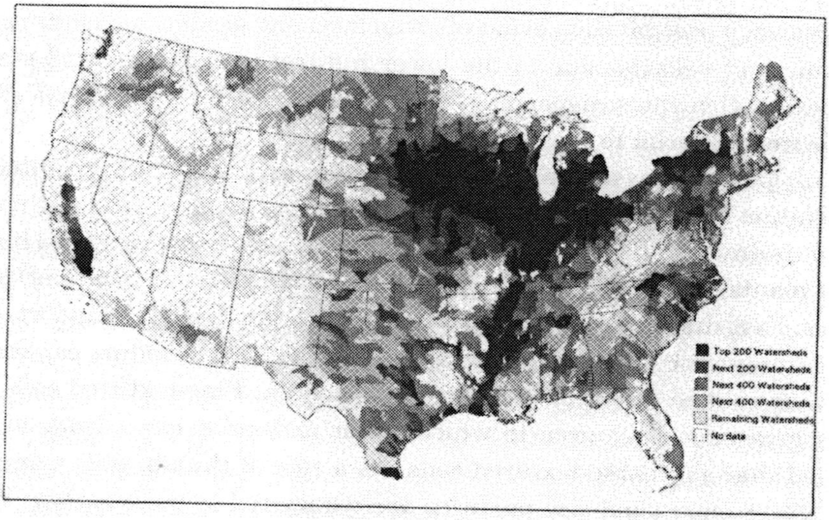

Figure 1.2 Map showing estimated potential pesticide runoff for the conterminous United States, (pounds per watershed), 1996. (Courtesy of USDA, NRCS.)

depth, and the availability of plant nutrients, as well as degrading soil structure and altering the soil texture (Weesies et al., 1994).

It has proven somewhat difficult to accurately document the relationship between soil erosion and land productivity because both soil erosion and productivity rates are influenced by numerous conditions that vary temporally and spatially. For example, Olson et al. (1994) observed that the variables complicating the relationship between soil erosion and soil productivity include (1) landscape position and hillslope components, (2) surface and subsurface water flow, (3) natural versus artificial erosion-control treatments, (4) soil properties, and (5) past and present land management. In addition, climate variability from year to year affects both soil erosion and productivity during field studies, complicating data analyses and interpretations.

Soil erosion is an insidious process that attacks the most productive topsoil layer first and may cause decreasing productivity at imperceptible rates over extended periods. Thus, the decline in soil productivity often is masked by planting high-yield crop breeds and by increasing the applications of fertilizers and pesticides where the financial resources are available to permit these investments (Follett and Stewart, 1985). Before the widespread use of commercial fertilizers, loss of topsoil reduced yields 50% or more compared to yields from soils with little topsoil loss (Weesies et al., 1994). Extensive analyses revealed that the reduction in crop yield depends on soil and climate characteristics as well as fertilization rates. Soil erosion affects farm-business profitability, in both the short and long term.

The cumulative effects of high erosion rates may have national and regional consequences. Since 1982, cropland in the United States, declined by about 11 million acres (4.5 million hectares) or 2.6%, pastureland declined by almost 12 million acres (4.9 million hectares) or 9%, and rangeland declined by about 11 million acres (4.5 million hectares) or 2.6%, while forestland increased by 3.6 million acres (1.5 million hectares) or 0.9% (USDA, NRCS, 1997b, revised 2000). Former croplands in the northeastern and southern states now support forests. Many acres (hectares) of Mississippi River bottomland forests and Great Plains grasslands are now croplands (USDA, NRCS, 1997a). In fact, there are several reasons for these land-use changes, including changes in agrobusiness economics, soil depletion due to past erosion, and the influences of erosion-control programs, in addition to competing land uses, especially urban and suburban development.

In an effort to protect soils vulnerable to water and wind erosion, the Conservation Reserve Program (CRP) was created in 1985. As of 1997, 32.7 million acres (13.2 million hectares) were enrolled in the CRP program and taken from production. The estimated average *annual* sheet and rill erosion from cultivated cropland was 3.1 tons/acre (7 metric tons/ha) per *year*, while the sheet and rill erosion from CRP land was 0.4 ton/acre (0.9 metric ton/ha) per year. Estimated average annual wind erosion from cultivated cropland was 2.5 tons/acre (5.6 metric tons/ha) per year, while wind erosion from CRP land was 0.3 ton/acre (0.7 metric ton/ha) per year (USDA, NRCS, 1997b).

Conservation reserve programs that retire erosion-vulnerable lands from production are feasible in the United States, due to the abundance of fertile croplands coupled with the economic and technical resources to maximize productivity on other lands. Today, each acre (hectare) of cropland produces nearly three times as much food and fiber as that which was produced by that acre (hectare) in 1935 (USDA, NRCS, 1997a). As a result, the demands of the U.S. marketplace are satisfied at prices lower than those for other industrial countries. The question is whether or not high levels of productivity are sustainable through future generations. There are about 108 million acres (43.7 million hectares) of cropland in the United States where the annual soil-erosion rate exceeds the soil-loss tolerance rate (USDA, NRCS, 1997b). Figure 1.3 shows the spatial variability of cropland erosion in the United States.

The situation, however, is very different in other parts of the world. There is some evidence indicating that the world's per capita food supply has declined during the past 10 years and continues to decrease, due to the loss of land productivity resulting from excessive soil erosion combined with population growth (Pimental et al., 1995). Where land ownership is not legally established, other lands are affordable, or simply as a matter of family survival, farmers may move to new lands when the expected returns from the new land exceed those from the existing land under cultivation (Barbier and Bishop, 1995). Shifting cultivation, itself, is not the problem; in fact, it is somewhat analogous to crop rotation. Intensive land use for extended periods prior to abandonment, including prolonged intense clean-tilled crop production, and the failure to protect the land with a vegetation cover while unused, can lead to severely degraded lands that require very long periods for recovery.

In some countries, government subsidies for fertilizers, pesticides, and irrigation facilitate the realization of productivity goals, at least for the short term. The focus of this volume is erosion processes, estimation, measurement, and control rather than a review of soil-erosion problems around the world. Information concerning soil erosion in other countries can be found in Morgan (1986), Troeh et al. (1991), and numerous issues of the *Journal of Soil and Water Conservation*, among other sources.

Land disturbance resulting from mining, residential, commercial, and highway construction can accelerate erosion rates by two or more orders of magnitude, depending on the pre-disturbance land use (Toy and Hadley, 1987). During a five-month period, 0.3, 2.0, and 0.4 ton/acre (0.6, 2.4, and 1.0 metric ton/ha, respectively) of sediment was trapped in weirs placed on three mined watersheds, while during the same period no sediment was trapped in the weirs placed on nearby unmined watersheds (Curtis, 1971). Even recreational land uses that disturb the vegetation cover and the underlying soil, including off-road vehicle traffic, horseback riding and hiking, can lead to rill and gully erosion, degrade the aesthetic appeal of the land, increase maintenance costs, and increase the sediment delivered to nearby streams.

The technology to control erosion rates carries a price tag that usually increases production costs on the farm, increases land-reclamation costs at the mine, and increases water- and sediment-management costs at the construction site; all of which reduce profitability. Historically, many

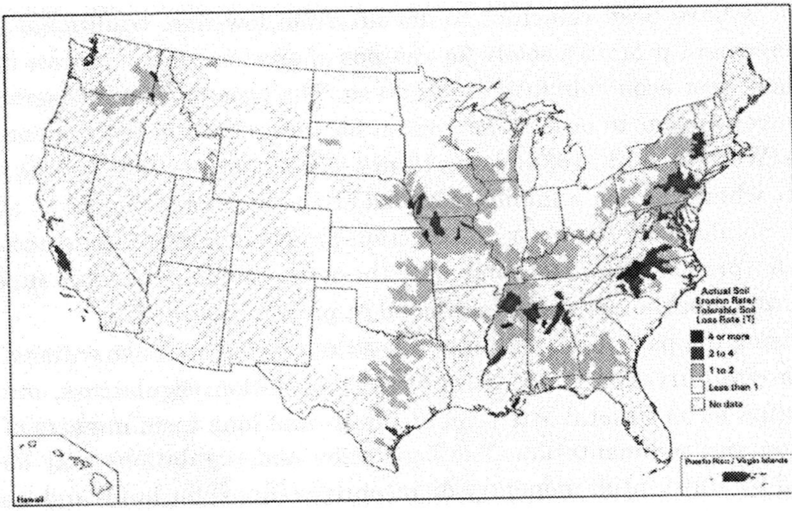

(a)

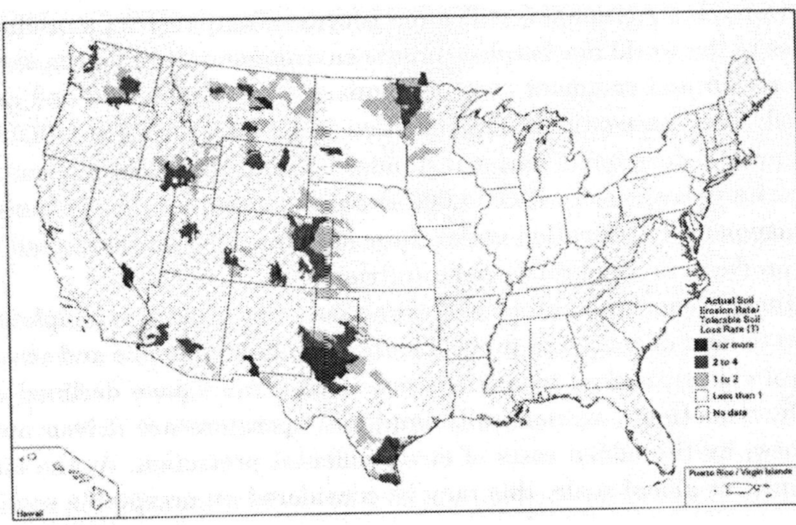

(b)

Figure 1.3 Maps showing (*a*) estimated average annual soil loss by water on cultivated cropland as a proportion of the tolerable rate (*T*), 1997, and (*b*) estimated average annual soil loss by wind on cultivated cropland as a proportion of the tolerable rate (*T*), 1997. (Courtesy of USDA, NRCS.)

farmers have been reluctant to deviate from low-risk, traditional land-management practices solely for reasons of erosion control, unless it was in their best economic interest to do so. The economics of conservation practices appear to be a major reason for their acceptance or nonacceptance (Weesies et al., 1994). Long-term erosion-control goals become irrelevant when there is a family to be fed or the mortgage banker is at the door. Similarly, the mine or construction-project manager is held accountable for project costs, and commonly there are company owners, supervisors, and stockholders very interested in project profitability.

During the past three decades, new economic factors have entered into the *profitability equation*. Environmental legislation, regulations, and certification focus societal attention on short- and long-term impacts of soil erosion and sedimentation. The legislation and regulations may be enforced by substantial monetary disincentives (fines) for land- and water-quality degradation or monetary incentives to control erosion and sedimentation in the form of participation in government subsidy or insurance programs. International certification programs may restrict a producer's access to the world marketplace unless environmental standards, including erosion and sediment control standards, are established and maintained. The International Organization for Standardization (ISO) is a worldwide federation of national standards organizations from about 140 countries (www.iso.ch). ISO 14,000 standards pertain to environmental management. Certification under these standards is required to sell specific products in some European countries.

Although legislation and certification does not provide a complete solution to land degradation problems, they can link economic and erosion-control objectives, and in many cases, erosion rates have declined as a result. Sometimes, economically marginal operators are driven out of business by the added costs of environmental protection. At the state, national, or global scale, this may be considered an acceptable sacrifice for the "greater good." At the personal or family scale, it seems like a high price to pay for a small increment of environmental protection.

Wind Erosion

Although discussions of erosion often focus on water processes, wind erosion deserves specific attention. Wind erosion cannot be dismissed as a problem of the American Dust Bowl during the 1930s (Figure 1.4). Wind erosion continues to be a problem in many parts of the arid and semiarid

Figure 1.4 "Black roller" moving across the plains carrying soil blown from unprotected farmland during the Dust Bowl days of the 1930s. (Courtesy of USDA, NRCS.)

world, including much of North Africa and the Near East; parts of eastern, central, and southern Asia; the Siberian Plains; Australia, northwestern China; southern South America; and North America (Wind Erosion Research Unit, 2001). (www.weru.ksu.edu/nrcs). In the United States, wind erosion is the dominant soil problem on about 74 million acres (30 million hectares). Wind erosion causes soil-texture changes because fine particles are removed, decreases soil depth and fertility, decreases land productivity, causes abrasion of plants, automobiles, and houses, causes sedimentation in ditches and on roadways, reduces visibility along roadways, and decreases water quality. In addition, research indicates that exposure to high aerosol concentrations of particles less than *10 μm* in diameter (PM_{10}) contributes to respiratory problems in humans. Urban areas on the Columbia Plateau of eastern Washington, northern Oregon, and the Idaho Panhandle have exceeded the PM_{10} standard established under the Clean Air Act of 1990 on numerous occasions since measurements started in 1985, with several occurrences on days of obvious wind erosion on agricultural lands (Saxton et al., 1997). Although somewhat speculative, Pimental et al. (1995) estimated total wind-erosion costs at about $9.9 billion per year for the United States—billions of dollars, each year, in one country. A comprehensive assessment of soil erosion and the development of erosion-control plans in any area requires consideration of both wind and water erosion.

Sedimentation

The sediment resulting from erosion is responsible for off-site environmental damage due to (1) sediment movement, (2) sediment storage, (3) chemicals adsorbed to sediments, and (4) the response of biota to both sediments and chemicals (Osterkamp et al., 1998). Sediment is, by far, the most common water pollutant in the United States as well as in other parts of the world. This sediment increases road-ditch maintenance costs, decreases the storage capacity and life expectancy of reservoirs, increases the cost of channel maintenance for navigation, decreases the recreational values of waterbodies, increases flood damage, and increases water treatment costs. The chemicals transported on sediment reflect regional land uses, such as fertilizers and pesticides in dominantly agricultural areas (Figure 1.2), and processing chemicals, heavy metals, and other manufacturing wastes in dominantly industrial areas. The cost of water pollution by sediment in the United States was estimated at about $16 billion per year in one recent study (Osterkamp et al., 1998), but this value was considered to be quite conservative. Another study set the total cost of erosion and sedimentation by wind and water at more than $44 billion per year (Pimental et al., 1995), but this value was considered to be too high (Crosson, 1995). The point is that the cost of erosion and sedimentation is measured in billions of dollars, each year, for the United States alone.

SOCIAL SIGNIFICANCE OF EROSION

Accelerated erosion due to human activities may first have occurred with the burning of vegetation to drive game animals, then intensified and spread geographically during the agricultural revolution, and again intensified and spread with technological developments that facilitated tillage and earthwork, including tractors, plows, disks and harrows, bulldozers, motor scrapers, road graders, and front loaders. Any activity that disrupts the vegetation cover on the land usually results in accelerated erosion rates.

An interesting history of agriculture compiled by Troeh et al. (1991) illustrates the significance of soil erosion in early civilizations. Although the actual first agricultural site may never be discovered, the ancient village of Jarmo in northern Iraq generally is considered to be the earliest (11,000 B.C.) location of cultivated agriculture. The villages in this area were situated on uplands with fertile, friable, and easily tilled silt-loam soils. It is logical to suspect that soil erosion was a problem on these sites and on these soils once the natural vegetation cover was disturbed. A common scenario emerges from the discussions by Troeh et al. (1991) concerning agriculture and soil erosion in Mesopotamia, Palestine, Phoenicia, Syria, Greece, and Italy. It appears that agriculture was first practiced on the floodplains, bottomlands, and other gently rolling, low-lying areas where erosion potentials were low. With increasing populations, the demand for food increased, causing agriculture and grazing to move up the valleysides onto steeper and more erodible hillslopes. Accelerated erosion on the hillslopes caused sedimentation on the fields and in the towns below and eventually reduced agricultural productivity. With increasing populations, the demand for fuels also increased, resulting in deforestation of hillslopes, again causing accelerated erosion rates over ever-wider areas. It seems that the citizens of Babylon lost the battle with sediment eroded from the surrounding hillslopes thousands of years ago. There undoubtedly are various factors contributing to the demise of ancient civilizations. Many agricultural and soil scientists believe that soil erosion often was a major factor, with declining agricultural productivity and increasing field and irrigation-system maintenance requirements.

Soil Erosion in the United States

In the United States, soil-erosion problems were noted during Colonial times as European farming practices were employed under North American environmental conditions. In the years that followed, advancements in agricultural practices frequently resulted in high erosion rates. New and sometimes highly erodible lands were put under the plow (Heimlich, 1985). By 1930, soil erosion by water and wind was recognized as a national problem requiring intervention at the federal level. The Soil Erosion Service was created by Congress on July 17, 1933, as a temporary public-works program that provided much-needed employment during the Great Depression, as well as the personnel for numerous erosion-control projects. In 1935, Congress transferred the agency to the U.S. Department of Agriculture and renamed it the Soil Conservation Service, later to become the National Resources Conservation Service in 1995.

Coal has been a major fuel source throughout the history of the United States. In Appalachia and other regions, coal was mined without much regard to environmental degradation until state laws to regulate the industry were enacted beginning about 1930. The effectiveness of these state laws in protecting the environment varied widely, and all too often, erosion transported sediment and adsorbed contaminants to stream channels, polluting the waters and destroying the aquatic habitats. The Office of Surface Mining Reclamation and Enforcement was created by Congress on August 3, 1977 to control surface coal mining operations and insure the reclamation of lands disturbed by these operations.

Hardrock mining activities can cause the same environmental problems as coal mining. In addition, the tailings from ore processing often are highly erodible and contain heavy metals and processing chemicals that can cause serious water-pollution problems if transported into and through drainage systems. Environmental regulation of hardrock mining activities has received considerable attention in recent years (National Research Council, 1999).

As the population of the United States increased and shifted from rural to urban areas, residential, commercial, and highway construction increased dramatically. These projects also disturbed the natural land surface, increased runoff rates and volumes by rendering a proportion of the surface impervious, and at least temporarily increased erosion and sedimentation rates in streams and rivers (Wolman and Schick, 1967). Montgomery County, Maryland was among the first U.S. counties, in 1965, to enact an ordinance requiring sediment control on all new residential construction sites.

The environmental movement of the 1960s and 1970s called attention to air, land, and water degradation in all forms. Soil erosion again was recognized as a serious environmental problem, caused not just by agriculture but by virtually all land-disturbing activities. Sediment was recognized as the primary pollutant in the nation's waterways. Laws were passed at federal, state, and local levels, followed by regulations and guidelines, to control soil erosion and sediment delivery to streams and rivers. A few examples of federal legislation that seek directly or indirectly to control soil erosion include:

- National Environmental Policy Act of 1969
- Clean Water Act of 1972
- Soil and Water Conservation Act of 1977
- Surface Mining Control and Reclamation Act of 1977
- Food Security Act of 1985 ("Farm Bill")
- Clean Air Act of 1990

For the first time in U.S. history, agricultural policy linked eligibility for federal farm-program benefits to land stewardship through the 1985 Farm Bill. This act required that farmers practice soil conservation in return for commodity price supports, crop insurance, farm loans, and other program benefits (Figure 1.5). Many other countries experience similar soil-erosion problems (Troeh et al., 1991) and have enacted laws and programs in efforts to control erosion.

State and local laws tend to parallel federal laws but emphasize special state and local environmental conditions or concerns. State laws usually must be at least as stringent as federal laws in order for a state to assume

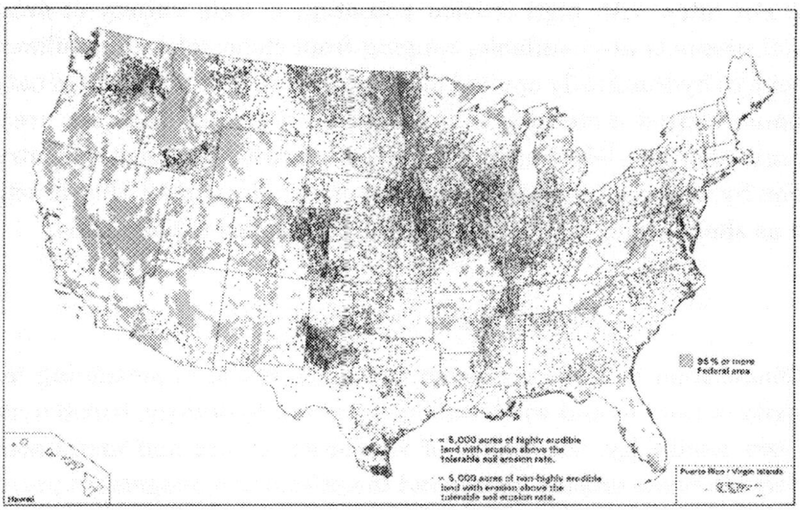

Figure 1.5 Map showing estimated excessive water and wind erosion on cropland, 1997. (Courtesy of USDA, NRCS.)

regulatory authority over activities causing land disturbance, accelerated erosion, and sedimentation.

Soil Conservation and Erosion Control

Many early civilizations recognized the relationship between accelerated erosion in their fields and the decline of agricultural productivity. Terraces appear to be among the first erosion-control practices, or perhaps the most persistent on the landscape. The Chinese, Phoenicians, Greeks, Romans, Incas, and Mayans were among those who constructed elaborate terrace systems. Vegetative cover and management have long been recognized as effective means of reducing erosion rates. Contour plowing and stripcropping have been regarded as effective erosion-control practices for many years. Crop-rotation sequences included years with complete vegetative cover on the field, such as four years of hay, two years of corn, followed by one year of wheat. The addition of manures and "night-soil" (human waste) have the combined effect of increasing productivity and reducing erosion rates. Standard agricultural conservation practices are a part of the reclamation of mine and construction sites as well.

In recent decades, erosion control has become a sophisticated science. Conservation tillage, especially no-till agriculture, reduces soil disturbance and leaves on the surface plant residues from the preceding year's crop. For sites with high erosion potential, a wide variety of erosion-control products are available, ranging from chemical soil stabilizers or binders, to hydraulically applied mulches, to nettings and mats of natural or manufactured materials, to interlocking concrete blocks for areas of concentrated flow. Most of the practices and products for the control of erosion by water also effectively control erosion by wind. Other practices, such as shelterbelts, are used specifically to reduce wind erosion.

SOIL-EROSION RESEARCH

The foundation of modern erosion control is research pertaining to the erosivity of rainfall and wind, hillslope and soil hydrology, runoff hydraulics, soil erodibility, the ability of vegetation covers and manufactured products to resist erosive forces, and the effects of management practices on both forces and resistances. This research must be conducted at various temporal and spatial scales. Temporal scales range from changes in erosion processes during a precipitation event to the potential influence of long-term climate change on erosion processes. The erosion rate and volume of sediment produced during a single design storm is important in the planning and construction of sediment basins, whereas average annual erosion rate usually is the basis for the selection of conservation practices on agricultural fields. Spatial scales range from erosion on areas no larger than a square yard (square meter) on ridges and hillslopes to the sediment discharged from entire watersheds.

The first formal erosion measurements in the United States were taken in 1912 on overgrazed rangelands in central Utah. In 1917, erosion plots were established at the University of Missouri–Columbia that are still in use today. Since the 1950s, rainfall simulation has been widely used to study erosion processes. Since the late 1960s to the present, much of the research focused on the mathematical modeling of erosion processes and rates using the accumulated field and laboratory data (Laflen et al., 1991a; Renard et al., 1991).

Formal erosion research programs at the federal level began in 1929 with the establishment of 10 soil-erosion experiment stations. In 1935, the Soil Conservation Service was formed, primarily to facilitate the transfer of erosion-control technologies from the laboratories and experimental plots to the farmer's field. In 1946, the Wind Erosion Project (now Wind Erosion Research Unit) was funded by the U.S. Congress and located at Manhattan, Kansas in cooperation with Kansas State University. In 1954, the Agricultural Research Service was created to coordinate USDA research, including soil-erosion research, and to provide technical support for the Soil Conservation Service. The National Runoff and Soil Loss Data Center, also was established in 1954 at Purdue University to assemble and consolidate for further analyses, all of the available runoff and erosion data collected throughout the United States. The National Sedimentation Laboratory was built in 1958 to study sediment transportation in streams and lake sedimentation; the mission was later expanded to include upland soil erosion and sediment transportation from agricultural-size watersheds.

Over the years, numerous university professors, their students, and government agencies throughout the world have contributed significantly to soil-erosion research, yet there is still much to be done, because erosion by water and wind have proven to be highly complex processes with rates varying in response to highly complex factors, such as weather, soil, vegetation, and land management. Today, it is possible to search the Internet for individuals, organizations, and government agencies involved in erosion research to learn of recent and current projects worldwide. Frequently, the latest reports and software can be downloaded from these Web sites. The Internet has revolutionized timely erosion-technology transfer (Appendix C).

TERMINOLOGY OF EROSION

Several scientific disciplines study erosion from somewhat different perspectives, for different reasons, and in different parts of the world. Over the years, terminology has emerged to describe erosion and related processes. Unfortunately, the terms are not employed consistently. Common terms and the definitions used in this book are discussed in Chapter 3. Types of erosion are differentiated on the basis of *location* (hillslope and channel erosion) and *agent* (water, wind, ice). Erosion and sedimentation processes include *detachment, entrainment, transportation*, and *deposition* of soil and other earth materials and may be described by traditional terms (e.g., rainsplash, sheet and rill erosion) and contemporary terms (interrill and rill erosion). *Soil loss* refers to the sediment from the eroding portion of a hillslope where overland flow occurs. *Deflation* is another term for wind erosion. *Sediment yield* and *sediment delivery* express the rate or amount of sediment transported to a point of measurement, at the base of a hillslope, the boundary of a field, in a stream channel, or at the mouth of a watershed. The term *denudation* usually pertains to large areas such as landscapes and drainage basins and refers to the average decrease in land-surface elevation during long periods of time, ignoring the variability of erosion rates within the large area. Soil scientists define the characteristics of a "soil" carefully and specifically. In this book, the word *soil* often is used in a broad context to refer to any soil-like material subjected to erosion processes, sometimes a true soil, sometimes a manufactured soil (minesoil), sometimes the alluvium of a floodplain. Students, instructors, scientists, and erosion-control specialists should exercise care in using these terms, in accordance with the conventions of their discipline, but remain aware that connotations associated with the terminology may differ among disciplines.

Natural (Geologic) Erosion

Erosion processes have operated for millions of years, as evidenced by the prevalence and thickness of sedimentary rocks at the Earth's surface. Natural erosion rates prevail under natural or undisturbed environmental conditions. Such erosion rates also are called *geologic rates*. Under natural erosion rates, soil properties and soil profiles usually develop to approach an equilibrium condition (Nikiforoff, 1942). A fully developed soil profile is sometimes taken as evidence that erosion is occurring at a natural rate. It is possible, however, for high erosion rates to occur in response to natural events such as fires or plant disease, extraordinary rainstorms, or congregations of wildlife that disturb vegetation covers and compact soils.

Accelerated (Anthropic, Human-Induced) Erosion

Under disturbed environmental conditions, water and wind erosion rates may be accelerated by several orders of magnitude. Such erosion rates also are known as *anthropic*, or *human-induced rates*. Soil formation generally cannot keep pace with soil removal. A degraded soil profile, with the B- or C- horizon at the surface, often is taken as evidence that erosion is occurring, or has occurred, at an accelerated rate. The land may be dissected by rills and gullies. Geologic strata may become exposed at the surface. Natural and cultivated vegetation productivity decrease sharply. Sediment is washed into ponds, lakes, and reservoirs or accumulates in channels to be transported by high flows. Agricultural chemicals or toxic materials are adsorbed and transported on the sediments. High sediment concentrations and chemical loads may adversely affect aquatic flora and fauna.

Usually, accelerated erosion is the result of human activities that disturb vegetation covers, expose soils, and increase slope steepness. Hence, accelerated erosion and its consequences largely are avoidable and manageable with proper planning and scheduling prior to land disturbance; implementation of water-, erosion-, and sediment-control practices during active disturbance; and land reclamation following disturbance. Since 1982, erosion on cropland and CRP land in the United States has decreased by about 38% (USDA, NRCS, 1997b).

DEVELOPMENT OF LANDSCAPES: A CONTEXT FOR EROSION

The landscapes of the Earth are the product of both internal and external processes. Internal processes originate within the Earth, add to the landmass above sea level; raise landmasses, increasing their potential energy;

and create surfaces upon which external processes operate. The enormous forces generated by these internal processes are manifested in earthquakes, volcanic eruptions, and folding and faulting of geologic strata. Internal processes are responsible for the large-scale (large-area) features of the Earth's surface, the positions of the ocean basins and continents, the rise and fall of broad structural mountain ranges and basins, and the deformation and tilting of strata within ranges.

External processes originate at or near the Earth's surface and work to level it by lowering the higher elevations and raising the lower elevations. There are four types of external processes: (1) rock weathering, (2) mass movement, (3) erosion, and (4) deposition. Rock-weathering processes are separated into two categories based on the changes in rocks and minerals. Physical-weathering processes result in the disintegration of rocks without changes in mineral chemistry. Chemical-weathering processes result in the decomposition of the minerals composing rocks.

Weathering processes substantially reduce the strength of rocks. The physical and chemical bonds between particles and compounds are greatly weakened. Gravitational forces produce shear stresses within the weathering mass. Eventually, the shear stresses exceed strengths, and mass movement occurs. Mass movements often produce hummocky, heterogeneous masses of earth and rock material near the downslope termini of hillslopes, referred to as *colluvium*. Lowering of the land surface by mass-movement processes is known as *mass wasting*.

Water, wind, or ice passing across land surfaces generate shear stresses within earth materials. As defined previously, erosion and sedimentation is the detachment, entrainment, transportation, deposition of soil and other earth materials. Sediment is the product of erosion.

The agents of erosion move sediment until the transport capacity of the flow is reached. Eventually, the transport capacity decreases, usually due to decreasing flow velocity. As transport capacity decreases, sediment in excess of the transport capacity is deposited. Sediments eroded, transported, sorted by particle size and density, and deposited by flowing water are known as *alluvium*. Only about 10% of the sediment eroded by water completes the journey to the sea. The remaining 90% of the sediment is deposited on hillslopes, on floodplains, and in lakes and reservoirs (Mead and Parker, 1985). Figure 1.6 shows the spatial variability in the amount of sediment delivered to streams in the United States.

Theoretically, land surfaces tend to become broad, virtually featureless plains. But usually, this is not what we see because internal and external

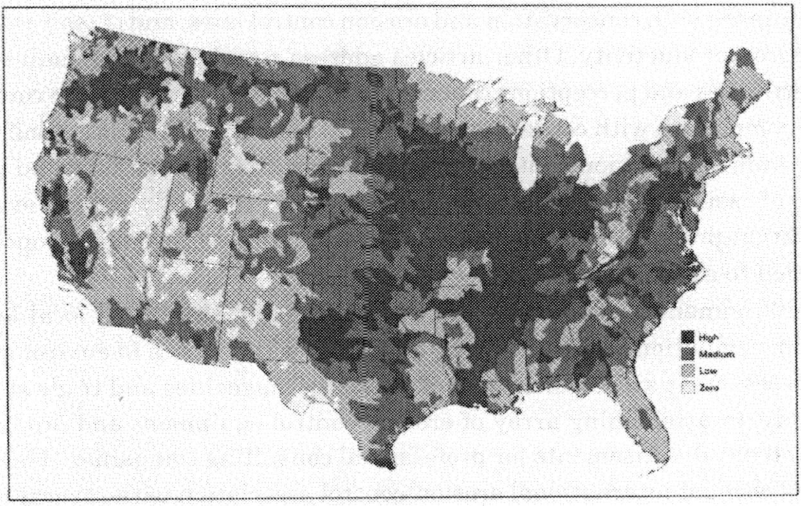

Figure 1.6 Map showing estimated sediment delivered to streams (tons), 1996. (Courtesy of USDA, NRCS.)

processes are constantly in action, varying in time and space. Topography is the current status of the interaction between internal and external processes. Most landscapes reflect the work of both past and present processes.

Erosion, then, is only one process group shaping the Earth's surface. It is, however, a very significant process group because it affects human welfare directly and indirectly and responds to human activities with increasing or decreasing rates. Throughout the remainder of this book we focus on upland erosion by water and wind. Most of the discussion centers on hillslope erosion, with due consideration to erosion in small channels. We believe that *in situ* weathering and mass wasting due to gravitational forces are separate and distinct processes.

SUMMARY

Erosion includes physical, economic, and social dimensions. During the past 100 years, most research centered on the physical processes. Again in recent years, as during the 1930s, economic and social issues have been addressed in erosion research. The conservation journals of the past decade contain articles examining the economics of erosion and erosion control, including the (1) cost-benefit and economic risks associated with various practices, (2) on- and off-site erosion costs, (3) economic impact of compliance with conservation and erosion control laws, and (4) soil erosion and crop productivity. Other articles address social issues, including the (1) attitudes and perceptions concerning soil conservation, erosion control, and compliance with conservation laws, (2) land-use planning for soil conservation and erosion control, (3) translating science into laws and policies, (4) soil-conservation education programs, (5) family factors in conservation-practice adoption, and (6) total-quality-management concepts applied to erosion control.

Environmental awareness, backed by national, state, and local legislation, regulation, and guidelines, created new industries in environmental assessment and erosion control. Industry magazines and trade shows display an astonishing array of erosion-control equipment and products, as well as advertisements for professional consulting companies. There is a professional international erosion-control association with several hundred members. Several countries have professional land-reclamation societies. Professional certification is available for erosion-and sediment-control experts, and the possibility of interdisciplinary, university-level, erosion- and sediment-control degree programs have been discussed by a professional erosion-control association. Erosion control is big business that translates into employment opportunities.

We take a somewhat nontraditional approach for an introductory soil-erosion textbook. First, the central concept is that the soil-erosion rate is a function of the soil-particle detachment rate and the soil-particle transportation rate. Hence, the types of erosion, the manner in which they operate, the principles of erosion prediction, and the principles of erosion control are examined in terms of particle detachment and transportation. Second, the interrill–rill concept provides the basis for the discussion of upland erosion processes. There are distinct differences in the erosion processes of interrill and rill areas although these processes are spatially linked. The interrill–rill concept has proven to be very useful in the development of erosion-prediction models and the development of erosion-control plans.

The underlying premise of the book is that a thorough understanding of erosion processes is the foundation for using and developing erosion-prediction models, selecting appropriate erosion measurement techniques for particular research designs, and selecting the best management practices for erosion control. This knowledge also provides a foundation for effective soil-erosion laws, regulations, and policies. Understanding the processes even helps us to understand the attitudes, perceptions, and economics surrounding soil conservation and erosion control. Imagine a small group of processes that qualifies as a major environmental problem worldwide. Imagine that you are a part of the solution.

SUGGESTED READINGS

Follett, R. F. and B. A. Stewart (eds.). 1985. *Soil Erosion and Crop Productivity*. American Society of Agronomy, Crop Science Society of America, Soil Science Society of America, Madison, WI.

Toy, T. J., and R. F. Hadley. 1987. *Geomorphology and Reclamation of Disturbed Lands*. Academic Press, San Diego, CA.

Troeh, F. R., J. A. Hobbs and R. L. Donahue. 1991. *Soil and Water Conservation*, 2nd ed. Prentice Hall, Upper Saddle River, NJ.

U.S. Department of Agriculture, Natural Resources Conservation Service. 1997. *A Geography of Hope*. U.S. Government Printing Office, Washington, DC.

2

Primary Factors Influencing Soil Erosion

Environmental conditions determine the types and rates of erosion operating in a particular area. These conditions consist of four primary components or factors: (1) climate, (2) topography, (3) soil, and (4) land cover and use. These factors influence both water and wind erosion, although in somewhat different ways. In this chapter, each factor is discussed individually and then in relation to the other factors. This discussion provides a foundation for subsequent chapters concerning erosion types, erosion processes, erosion-prediction technology, and erosion control.

WATER EROSION

Soil erosion by water is the dominant geomorphic process for much of Earth's land surface. Water erosion and sedimentation include the processes of detachment, entrainment, transport, and deposition of soil particles. The major forces driving these processes are shear stresses generated by raindrop impact and surface runoff over the land surface. Water erosion is a function of these forces applied to the soil by raindrop impact and surface runoff relative to the resistance of the soil to detachment. Once set in motion, soil particles are referred to as *sediment*. *Sediment delivery* is the amount of eroded material delivered to a particular location, such as from the eroding portions of a hillslope (soil loss) or the outlet of a watershed (sediment yield). The capacity for runoff transporting sediment is also related to shear stresses applied by runoff to the soil surface and the transportability of the sediment, which is related to the size and density of the sediment particles. If the sediment available for transport becomes greater than the transport capacity, deposition results in the accumulation of sediment on the soil surface. Therefore, environmental conditions determine erosion rates, sedimentation (deposition) rates, and sediment yield. The four environmental factors that determine water erosion and sedimentation are climate, soil, topography, and land use. Each factor operates both independently and interactively. In this chapter we provide an overview of the main effects of the four factors.

Rainfall erosivity, which is determined by climate, provides estimates of the forces applied to the soil to cause water erosion. Soil has properties that determine its inherent erodibility (susceptibility) to erosion. Topography, vegetation, and soil surface configuration modify the forces applied to the soil, and the presence of biological materials and management of the soil affect the soil's erodibility.

Climate

Climate refers to weather and the variability of weather conditions through time. Climate influences erosion directly and indirectly. Precipitation is the single most important climate variable affecting water erosion. Average annual precipitation varies throughout the world from 0.03 in. (0.08 cm) at Arica, Chile, to 460 in. on Mt. Waialeale, Hawaii. Figure 2.1 is a map of average annual precipitation for the continental United States.

Erosion by rainfall occurs from raindrops striking soil, and water flowing over the soil. Several variables could be used to describe the erosivity of rainfall. These variables include rainfall amount, kinetic energy, momentum, and intensity (Foster et al., 1982a, Wischmeier, 1959). Common observations show that the two important rainfall variables that determine storm erosivity are rainfall amount and rainfall intensity. *Rainfall intensity* provides a measure of erosion per unit rainfall, which multiplied by rainfall amount provides an estimate of total erosivity for the storm.

This simple relation is obvious for erosion by raindrop impact, and it also applies to erosion by surface runoff. Erosion by surface flow is related to both rate and amount of runoff. *Amount of runoff* is related to rainfall amount, less the amount of infiltration, and *peak runoff rate* is related to peak rainfall intensity, less infiltration rate. Therefore, the product of rainfall amount and rainfall intensity applies as a simple measure of rainfall erosivity for erosion by both raindrop impact and surface runoff.

Although rainfall amount and intensity capture much of rainfall erosivity, a further consideration related to *raindrop size* is needed. The

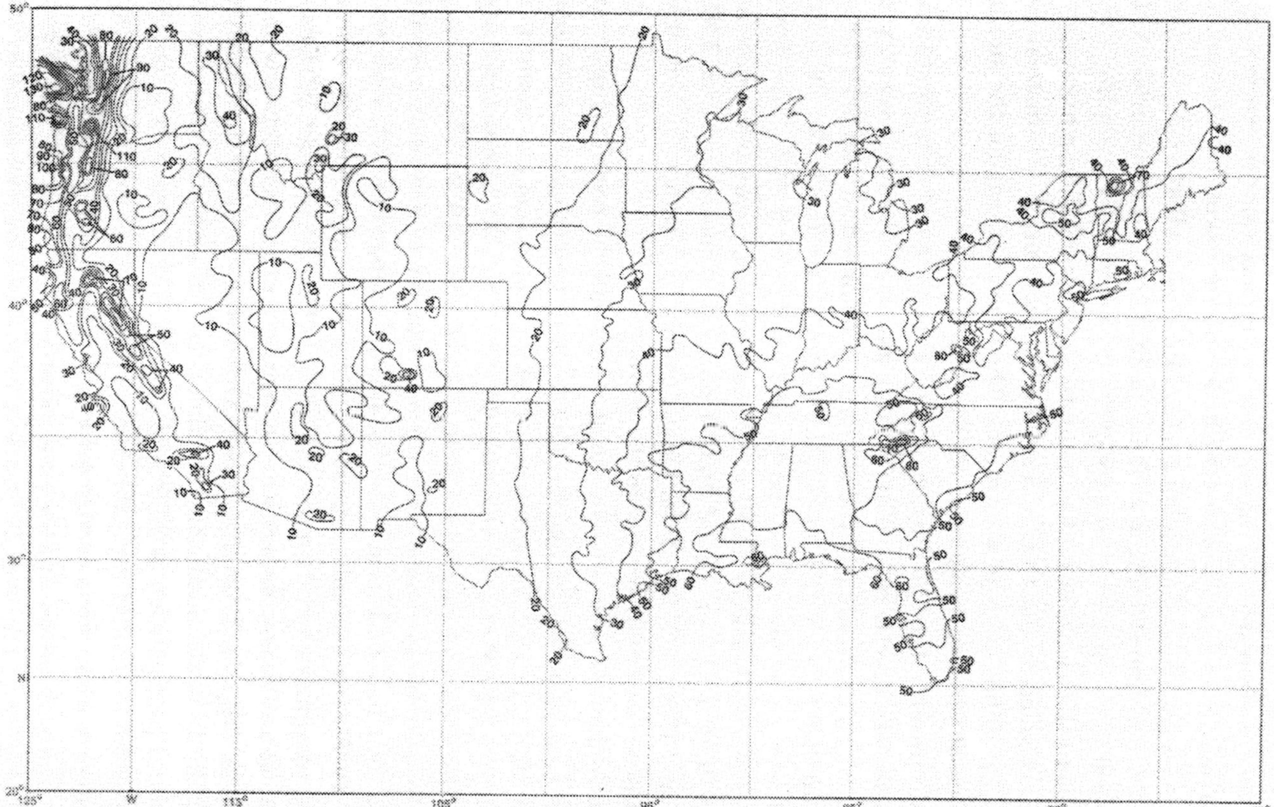

Figure 2.1 Average annual precipitation in the United States (in.). (To convert to millimeters, multiply by 25.4.) (Owenby et al., Precipitation Map, Climatography Of The U.S. No. 81—Supplement #3, NOAA".)

forces applied to the soil by the raindrop impact is related to drop size. A very small raindrop striking the soil at a low impact velocity exerts very low force on the soil and causes very little erosion regardless of rainfall amount and intensity. A variable that captures this effect is the *kinetic energy* of a raindrop striking a surface, which is one half of the product of the mass of the drop and the square of the impact velocity. Raindrop size and impact velocity are very closely related. A very small drop has a very low impact velocity and thus a very low impact energy. A rainstorm is made of up of thousands of waterdrops. The kinetic energy for the storm is the sum of the kinetic energies of individual raindrops.

Raindrops not intercepted by vegetation strike the soil at terminal velocity, which is directly related to raindrop diameter. Terminal velocity is reached when the weight of the raindrop is balanced by the drag force that is applied to the raindrop as it falls through air, much like the drag applied to a parachute. The total kinetic energy for a 30-in. (760-mm) annual precipitation occurring over 1 square mile (2.6 square kilometers) is equivalent to the energy of 10,000 tons (9100 metric tons) of TNT (Meyer and Renard, 1991). Total kinetic energy for a rainstorm is nearly proportional to rainfall amount in the storm.

The complexity of erosion requires that many of the relationships that describe erosion be determined by experimental research. Such is the case with rainfall erosivity. Experimental research show that the product of total kinetic energy of a storm times the maximum 30-minute intensity is a good measure of rainfall erosivity (Wischmeier, 1959). The average sum of kinetic energy of all storms in a year is an index of rainfall erosivity that can be calculated from rainfall data and mapped (Wischmeier and Smith, 1978). For example, erosivity in New Orleans, Louisiana, where average annual precipitation is about 100 in. (250 cm), is about 100 times erosivity at Las Vegas, Nevada where average annual precipitation is only about 4 in. (10 cm). Erosivity at New Orleans is about 10 time greater than erosivity in the north-central part of the United States where average annual precipitation is 30 in. (75 cm). Note that erosivity is not directly related to rainfall amount because of the effect of rainfall intensity. This range in erosivity is even greater in certain areas of the tropics.

Annual precipitation, and thus annual erosivity, vary greatly from year to year. For example, over 20 years of record at many locations in the United States, the erosivity in the year of maximum erosivity was about seven times that in the year of minimum erosivity (Wischmeier and

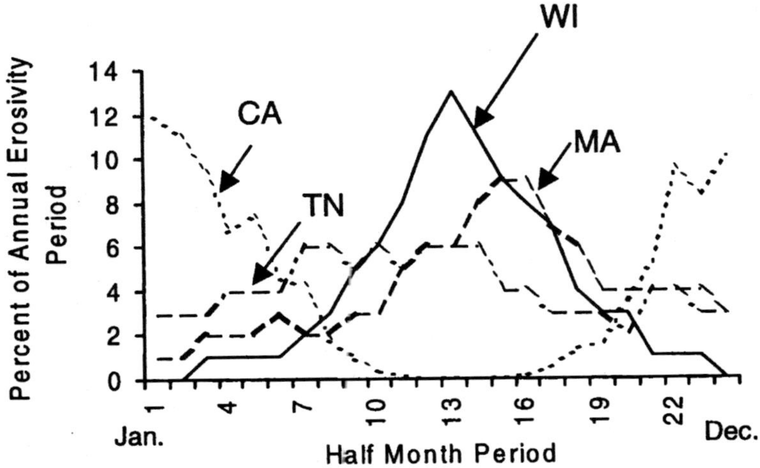

Figure 2.2 Variation of erosivity during the year for stations located in the United States.

Smith, 1978). Erosivity also varies within the year. Figure 2.2 illustrates the average variation in erosivity over a year for four locations in the United States. Much of the annual erosivity occurs during the winter and early spring months at the California location. A major peak in erosivity at an Arizona location occurs in the summer and a minor peak occurs in the winter. Erosivity is concentrated in the late spring and early summer in Wisconsin, while erosivity is concentrated in late summer and early fall in Massachusetts. Erosivity is uniformly distributed through the year in Tennessee, which illustrates how both rainfall amount and intensity affect erosivity for that location. Temporal erosivity is important because it interacts with variations in vegetation and soil conditions during the year to affect erosion significantly. Erosion is greatest when the peak period of erosivity corresponds to the period when the soil is most exposed to raindrop impact and surface runoff. Significant reduction in erosivity can be achieved by ensuring that these two periods of peak erosivity with peak susceptibility do not correspond.

Precipitation form, which is determined primarily by temperature, strongly influences precipitation erosivity. Snowfall is not erosive, whereas high-intensity rainfall is very erosive. Temperature also influences soil conditions, and the interaction of precipitation form and soil conditions can result in very low or very high erosion rates in different periods of the year. Snowfall and rainfall on frozen soil produces no erosion, but runoff from snowmelt and rainfall on thawing soils can produce very high erosion rates, as recorded in the northwestern United States (Renard et al., 1991, 1997).

Although rainfall amount and intensity are basic erosivity factors, runoff amount and rate must also be considered as erosivity factors for surface runoff. Infiltration, which is a function of antecedent soil moisture and soil sealing, must be estimated. Runoff occurs when the rate of precipitation onto the surface exceeds the infiltration rate into the surface. Thus, an understanding of hydrology is essential to an understanding of water erosion (Appendix B).

Vegetative cover provides surface protection from the forces of raindrop impact and surface runoff. Plant roots, organic matter, and the products of decomposition in the soil increase soil resistance to erosion. The climatic variables of precipitation, temperature, evapotranspiration, and in turn soil moisture influence vegetation (biomass) production and decomposition within the soil profile. These climatic variables affect vegetative growth and decomposition of plant materials left by the vegetation, which in turn greatly affect erosion. The amount of vegetation produced is related directly to rainfall amount. Vegetation, or parts of vegetation that die each year, such as leaves falling on the ground and roots sloughing, adds organic material to the soil surface and in the soil, which reduces soil susceptibility to erosion. The amount of organic material left after the vegetation depends on vegetative biomass that was produced and how much is left after other uses, such as harvest for fuel or animal feed. Organic material left on and in the soil is lost over time by decomposition. Decomposition occurs much more rapidly in locations where rainfall and temperature are highest, such as in the southeastern United States, especially during winter months.

Soil

In general terms, *soil* refers to loose material mantling the surface of Earth, as distinct from solid rock (Appendix A; Govers and Poesen, 1986). Soil serves many functions. It nourishes and supports growing plants. It is a construction material that supports buildings and roads. It also produces sediment by erosion that can fill reservoirs and water conveyance channels, be a pollutant itself, and carry adsorbed chemicals that degrade water quality in streams and lakes.

Soil evolves through time. Initially, bedrock weathers to unconsolidated mineral debris that serves as the parent material for pedogenesis (soil development). Frequently, sediment transported by water (alluvium) or wind (loess) provides the parent material. Soil development eventually reflects the combined influences of climate, topographic relief, organic activity, and time, such as different environmental conditions produce in different soil properties.

Some soils are inherently much more erodible than other soils (Agassi et al., 1996; Bryan, 1977; Foster et al., 1985; Foth & Turk, 1972). Highly erodible soils can be 10 times as susceptible to erosion as erosion-resistant soils. Other factors besides inherent soil properties affect soil erodibility (inverse of soil resistance to erosion) such as organic material in the soil. A standard reference condition is used to measure soil erodibility, where all effects have been removed except for the effect of inherent soil properties. One such reference condition is to prepare a plot where the soil is maintained in continuous fallow (no vegetation) that is periodically tilled up and down slope (Chapter 6). Sediment from the plot is collected and plotted as a function of erosivity (Figure 2.3). The slope of the line is an empirical measure of erodibility.

Soil texture is the single most important soil property in many applications, especially in erosion. Soils that are high in clay have low soil

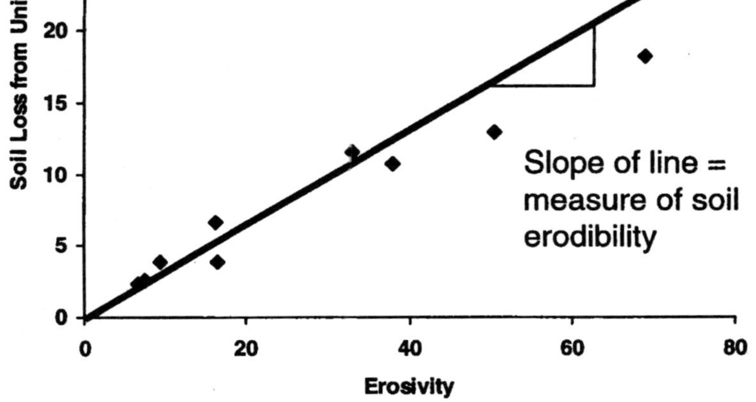

Figure 2.3 Soil-erodibility determination for standard test condition.

erodibility values because these soils are resistant to detachment. Soils that are high in sand also have low soil erodibility values, at least where the erosivity index is based on total kinetic energy (essentially rainfall amount) and rainfall intensity. These soils produce less sediment because of reduced runoff from high infiltration. Even though soil particles from sandy soils can readily be detached, not much runoff is available to detach and transport soil particles from sandy soils. Medium-textured soils, especially soils very high in silt, are the most erodible. These soils tend to produce increased runoff, soil particles are easily detached from these soils, and sediment eroded from these soils is easily transported. Permeability of the soil profile influences erodibility because of its influence on runoff. Restricting layers, such as a clay pan, near the soil surface can increase erodibility significantly, especially on sandy soils that are easily detached. Increases in runoff on these soils can result in significant increases in erosion. Other factors that affect erodibility include soil structure, which is the arrangement of soil aggregates within the soil, and the presence of binding agents of organic compounds and iron and aluminum oxides (Grissenger, 1966; Moldenhauer and Wischmeier, 1960; Wischmeier et al., 1971).

Soils vary in erodibility during a typical year. Soils are more erodible based on the product of rainfall energy and intensity during those months when soil moisture tends to be high. Increased soil moisture increases runoff amount and peak rate per unit of rainfall that produces increased erosion. Soil moisture is increased in those months when rainfall is high and temperature is low. Low temperature increases soil moisture because less water is evaporated from the soil at low temperature than at high temperature. Thus, soil erodibility tends to be higher during the later winter and early spring months than in the late summer months. Another reason that erodibility is reduced in late summer months is that higher temperatures during the summer months increases biological activity in the soil and produces organic compounds in the soil that are bonding agents. Wetting and drying cycles decrease erodibility and these cycles are reduced during the winter months (Kemper et al., 1985). Thus, erodibility in the spring is elevated. Soil that is thawing is highly erodible with respect to surface runoff (Van Klaveren and McCool, 1998). This condition occurs routinely in the Northwestern Wheat and Range Region in the Palouse area of the United States (Austin 1981). It also occurs, but less frequently, in late winter in other locations. When rainfall occurs on thawing soil, erosion by surface runoff can be especially severe.

Soil compaction and dispersion of fine soil particles caused by raindrop impact can cause a dense layer to develop at the soil surface (Bradford and Huang, 1992; Bradford et al., 1986; Sumner and Stewart, 1992). This layer, known as a *crust* when dry and as a *seal* when wet, greatly reduces infiltration rate and increases runoff, which in turn increases erosion.

Soils vary in their degree that they are susceptible to erosion by raindrop impact relative to their susceptibility to erosion by surface runoff. The case of runoff on thawing soils is an example of a situation where the soil is much more susceptible to erosion by runoff than by raindrop impact. A clay soil can be highly resistant to erosion by flow in comparison to a high silt soil that is easily eroded by surface runoff. Erosion by raindrop impact is uniform over the landscape, whereas erosion by runoff increases with distance along the runoff path (Meyer et al., 1975b). The effect of topography on erosion is greater where a soil is more susceptible to erosion by runoff than by raindrop impact.

Erosion varies over the landscape because soil properties vary over the landscape. For example, soil behavior at the top of hillslopes is very much determined by the parent material from which the soil is derived (Renard et al., 1997). In contrast, soils on lower parts of the hillslopes where deposition has occurred over centuries reflects the properties of a soil developed in depositional environment that is quite different from the soil development environment at the top of the hillslope. An important effect on soil development and the consequent soil properties is the variation of moisture regimes over the landscape. The lowest parts of the landscape and those areas in the hollows tend to be wetter than other parts of the landscape, which causes the soil properties to be different in those areas because of differences in the soil development environment.

Topography

Topography refers to the geometry of the land surface. The important geometric variables are slope length and steepness, shape in the profile view, and shape in the plan view. Uniform slopes are the simplest slopes. Uniform slopes are those where steepness doesn't change along the slope. Erosion increases along uniform slopes because of the accumulation of runoff along the slope. Figure 2.4*a* illustrates the common slope shapes. A simple plot of erosion along a uniform slope is shown in Figure 2.4*b*. Erosion from raindrop impact and surface runoff is combined into a single term where erosion varies as a function of the distance along the slope to

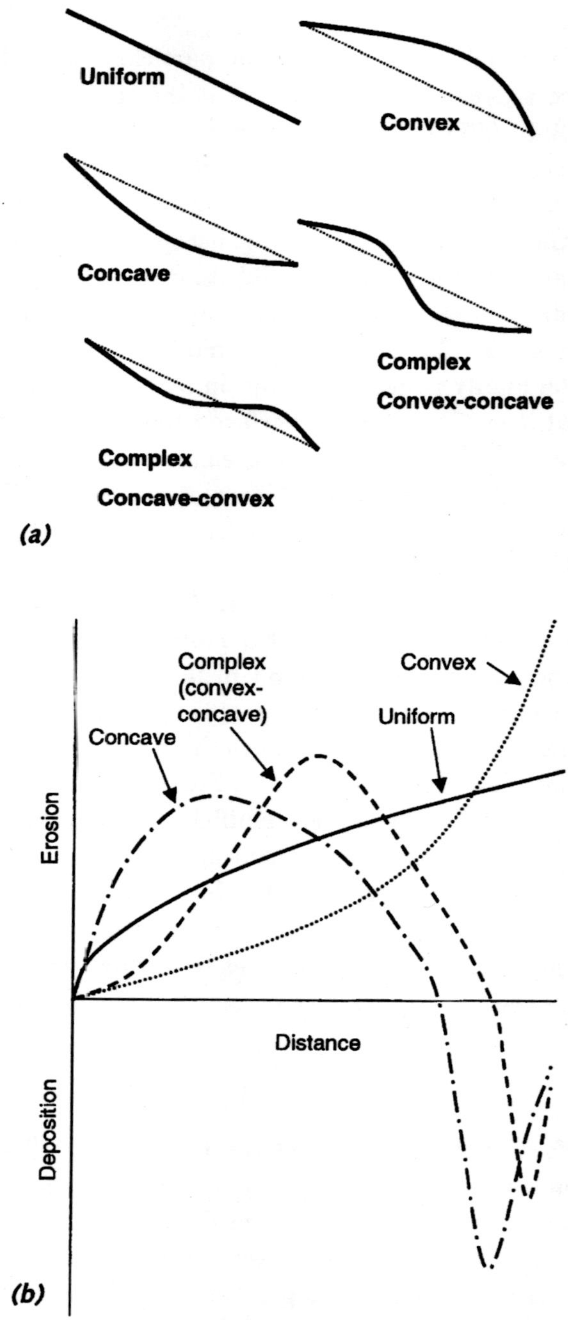

Figure 2.4 (*a*) Hillslope shapes for different topographic scenarios; (*b*) erosion and deposition for the hillslope shapes in (*a*).

the approximate 0.5 power. Sediment delivered from the end of the slope (soil loss) varies as a function of the length of the uniform slope to the same 0.5 power. The 0.5 power varies depending on how much erosion occurs by raindrop impact relative to how much erosion is caused by surface runoff. The power increases as the amount of erosion caused by surface runoff increases relative to the amount of erosion caused by raindrop impact, and conversely, the power decreases as the relative amount of erosion caused by raindrop impact increases (McCool et al., 1989).

Erosion is also related to the steepness of the uniform slope (McCool et al., 1987). As slope steepness increases, the increase in erosion is linear with the increase in steepness. More of the slope steepness effect comes from erosion by surface runoff than by raindrop impact because steepness has a greater effect on erosion by flow than by raindrop impact (Foster, 1982).

Thus, erosion at a location on a slope is a function of the distance from the surface runoff origin and the steepness at that location (Foster et al., 1977). If the location is far down the slope where much runoff has accumulated, the erosion rate will be high. For a given location, erosion will be proportional to the steepness at that location. This principle can be applied to nonuniform slopes that are common on natural landscapes. Figure 2.4*a* shows several common slope profiles. These profiles are actually flow paths that trace the water flow from its origin downslope to a concentrated flow area such as a terrace channel or ephemeral gully (Chapter 3). The path is perpendicular to the topography contour lines (lines of equal elevation) assuming that no ridges are present to redirect the runoff from its most direct downslope path. The flow path and its slope length end where the flow enters a concentrated flow area. Slope lengths range from shorter than 20 ft (5 m) on hillslopes highly dissected by flow concentrations to greater than 1000 ft (300 m) on gentle hillslopes without many flow concentrations.

A *nonuniform slope* is one where slope steepness varies along the flow path (Figure 2.4*a*). Of course, other properties, such as soil characteristics, can vary along the path, but only a variation of steepness is being considered here. A *convex slope* is one where steepness increases continuously along the slope. As described above, erosion at a location on the slope is related to distance from the slope beginning at the location in question and the slope steepness at the location. Figure 2.4*b* illustrates the erosion rate along a convex slope in relation to that for a uniform slope. Maximum erosion rate on the convex slope is much greater than maximum erosion rate on the uniform slope, and sediment delivered from the end of the convex slope is greater than that from the uniform slope. The combination of runoff indicated by the location on the slope and slope steepness is greater at the end of a convex slope than at the end of a uniform slope.

A *concave slope* is the converse of a convex slope. Runoff is least on the concave slope, where steepness is greatest (Figure 2.4). Maximum erosion rate is slightly less on a concave slope than on a uniform slope, and sediment yield from a concave slope is less than that from a uniform slope for a case where steepness of the concave slope doesn't flatten sufficiently to cause deposition.

Deposition occurs on a slope where the amount of sediment available for transport becomes greater than transport capacity (Chapter 4). Sediment available for transport at a location on a slope is related to the amount of sediment produced by erosion on the upper part of the slope. Transport capacity for a location on the slope is directly related to location on the slope, which determines runoff, and slope steepness. Thus, if the steepness of a concave slope is sufficiently flat at the lower end, much deposition can occur, which greatly reduces sediment delivery at the end of the slope. The complex slope with an upper convex segment and a lower concave segment behaves as a combination of the convex and concave slopes.

Figure 2.5 illustrates landscape shapes in the plan view. The landscape has three important parts. One part is the *valleysides*, side slopes where runoff is parallel along the flow path (location D). Discharge rate per unit width across the hillslope is the same all along the flow path. The second

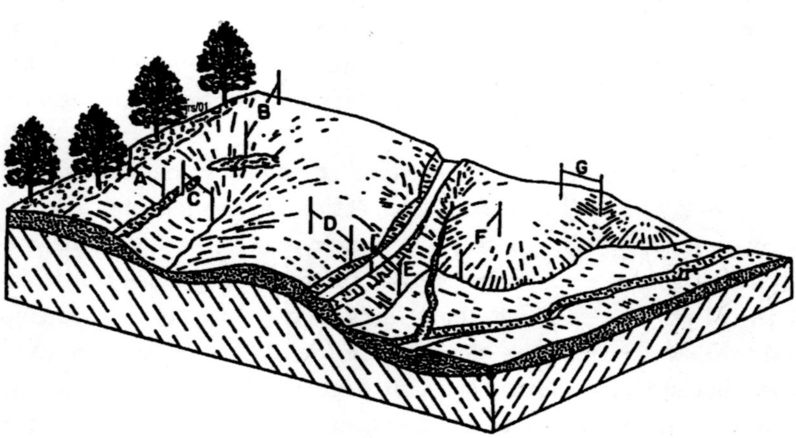

Figure 2.5 Typical hillslopes composing landscapes.

important part of the landscape in the plan view is on the *noses* or spur-ends of the hillslope (location F). On these areas, the flow diverges such that the discharge per unit width decreases in a downslope direction, and thus the erosion rate per unit width would decrease in a downslope direction. The third important location are the *hollows* that cause runoff to converge (location G). In these areas, the discharge per unit width increases downslope with a corresponding increase in erosion per unit width in the downslope direction. Concentrated flow areas form in the hollows and swales where runoff has converged, to where the dispersed flow around the hillslope has been collected into a single channel.

Topography also has an indirect effect on erosion in the same way that climate has an indirect effect on erosion by how it affects vegetation. Dimensionality in the landscape, especially in the plan view, causes soil moisture to be higher on lower parts of the landscape than on the upper parts of the landscape. For example, soil moisture tends to be higher at the base of hillslopes and in the hollows than on the upper hillslope portion and on the spur-end slopes (Weltz et al., 1998). Vegetation, biomass production, and organic matter on and in the soil tend to be greater for those areas where soil moisture is high, which contributes to the spatial variability in erosion over a landscape in addition to the simple effect of location on runoff and the effect of slope steepness on the erosivity of raindrop impact and surface flow (Foster, 1982).

Land Use

Land use has more effect on erosion than does any other single factor. *Land use* refers to both general land use and the management applied to that land. For example, land use can be a totally undisturbed area maintained in pristine condition by excluding all human activity. Another land use might be forestland, where disturbance is generally limited to logging and reseeding. Another land use is rangelands, where cattle are allowed to graze. Vegetation can be much greater where grazing is carefully controlled, in contrast to where overly intense grazing removes an extensive amount of the vegetation. Type of vegetation is determined primarily by climate and soil, but the amount of vegetation is determined by management on farmed and grazing land. Soil disturbance, which affects erosion in addition to vegetation, is also determined by management. During construction at an urban development site, filling of a landfill, or mining, all vegetation has normally been removed and soil is fully exposed to erosion.

After the severe disturbance, the land is reclaimed and restored by applying surface cover (mulch) and establishing vegetation to control erosion. Another land use is military training and recreational areas, where the degree of vegetation and soil disturbance is determined by the level of use. Thus, land use refers broadly to the vegetative cover, applied cover, and level of vegetative and soil management collectively, along with supporting practices, such as contouring, that are applied to the land to control erosion.

Erosion occurs in exactly the same way on all land uses. Erosion is related directly to the forces applied to the soil by the erosive agents of raindrop impact and surface runoff in relation to the resistance of the soil. Land use and land-use activities affect both the forces applied to the soil and the resistance of the soil to those forces. The following discussion focuses on three elements of land use: (1) the roles of vegetation, (2) the surface cover provided by plant residue or applied materials, and (3) mechanical soil disturbance (Renard et al., 1997; Weltz et al., 1998; Wischmeier, 1975).

Vegetation

Canopy Vegetative canopy is the aboveground part of the vegetation that intercepts raindrops but does not touch the soil surface to affect surface runoff. Dense canopies that cover a high percentage of the soil surface intercept a large proportion of the rainfall. Some of the intercepted rainwater is evaporated without reaching the ground to cause erosion. This interception can be significant in dense canopies having high leaf area per unit of ground area that runoff is greatly reduced for small to moderate rainstorms.

Some of the rainwater intercepted by the canopy flows along plant surfaces to reach the stems and in turn move as stemflow to reach the soil. Erosion by raindrop impact from this water is eliminated. Although this water becomes a part of surface runoff, its erosivity as runoff is reduced because of the travel time from the interception point on the plants at the soil surface.

Some of the rainwater intercepted by plants forms waterdrops that fall from the plant to the soil surface. If the bottom of the canopy is sufficiently close to the soil surface, erosivity of this water is much less than the erosivity of raindrops that are not intercepted. However, if the canopy bottom is sufficiently high, such as with some trees, the fall height of drops is sufficient for them to approach the erosivity of rainfall not intercepted by canopy (Chapman, 1948).

The degree that canopy affects erosion depends on the percent of the land surface covered by the canopy and the density of the canopy. Open space can exist between the outer perimeter of each plant and within the perimeter of each plant, which is space where raindrops can strike the soil directly without their erosivity being reduced in the canopy. Open space between plants is reduced by close plant spacing, and open space within the canopy perimeter is reduced by plants that produce much leaf surface area and by increased ground biomass production (U.S. Department of Agriculture, Forest Service, 1987). The reduction of erosivity by the canopy is reduced by plants that have their canopy bottom close to the ground.

Canopy varies during the year with natural growth processes. Canopy increases as plants come out of their dormant periods or after they have been seeded. After maturity, plants lose canopy by senescence as leaves droop and fall to the soil surface. Maximum erosion control is gained by choosing plants and managing them so that they have their maximum canopy during periods of maximum erosivity.

Plant canopy varies spatially as plant communities vary over the landscape as a result of soil properties, soil moisture, aspect, and other microclimate effects that vary over the landscape. Different plant types can occur at the same location, such as dense grass cover under a desert shrub. The grass canopy has a greater effect than the shrub canopy on reducing erosivity.

Plant Stems and Basal Area Plant stems are important as well as the basal area of the plant at the soil surface. Erosion is eliminated on that portion of the soil covered by plant basal area. The basal area cover can often amount to as much as 20 to 30% in dense vegetation. Dense plant stems slow runoff and reduce erosivity. Widely spaced plant stems can actually accelerate erosion around the stem in the same way that flow around a bridge pier causes local erosion. Overall, plant stems and basal area reduce erosion.

Some of the roots of perennial plants go through an annual cycle of growth and decay (root sloughing) similar to aboveground senescence. Up to 40% of the root biomass can be sloughed in a year (Reeder et al., 2001). The entire root biomass from an annual plant such as wheat is left in the soil after harvest. This root biomass provides soil biomass that decomposes to greatly reduce soil erodibility and significantly increase infiltration. Soil water can flow along live roots and in the channels left by decaying dead roots. These channels increase infiltration, reduce runoff, and decrease erosion.

Aboveground biomass is incorporated into the soil by tillage or other mechanical soil disturbances that mix plant materials into the soil. Biomass can be added to the soil in manure and with other organic waste materials. Earthworms and other organisms that live in the soil incorporate plant material in the upper 2 in. (50 mm) of the soil (West, 1990). Plant residue near the surface has a much greater effect than when buried deeply in the soil. Floral and faunal activity in the soil improves soil structure and reduces erosion.

Erosion is reduced as the amount of biomass in the soil is increased by vegetation type and production (Renard et al., 1997). Just as the aboveground biomass production varies with space, so does belowground biomass. The accumulation of organic material in the soil depends on the mass of organic material in the soil. The organic material amount that remains depends on decomposition rate. Decomposition depends on soil temperature and moisture, which are closely related to air temperature and rainfall.

Vegetative strips are sometimes placed on a hillslope or at the edge of streams to trap sediment before it leaves an on-site source area. These strips, known as *filter strips*, slow the runoff sufficiently to cause deposition, thereby reducing sediment yield from the area. Strips of dense vegetation placed up on the hillslope are known as *buffer strips*, and similarly these strips reduce erosion and retain sediment on the hillslope. Rotational stripcropping involves planting alternate strips of close-growing vegetation with strips of intensively tilled crops (Figure 2.6). The strips of close-growing vegetation and other crops are rotated sequentially so that all vegetation types eventually are grown on every strip. The width of all strips in the rotation system is usually the same. Stripcropping reduces the amount of sediment moving down the hillslope. The effectiveness of vegetative strips depends on plant density and the ability of the strip to slow runoff and cause deposition.

Ground Cover

Ground or *Surface cover* is material in direct contact with the soil that protects the soil from raindrop impact and slows surface runoff. Ground

Figure 2.6 Stripcropping in Columbia County, Washington.

cover can result from plant litter that accumulates from leaf fall, crop residue added with harvest, straw mulch, or manufactured materials added to the soil surface, naturally occurring rock, and live parts of plants touching the soil surface. Empirical research has shown that the effect of ground cover on erosion is related directly to the percent of the surface covered. The effect of ground cover, described by relationships such as those illustrated in Figure 2.7 varies among climate, topography, and soil conditions. For example, natural rock cover on a rangeland site is more effective at reducing erosion than is rock placed as mulch on a construc-

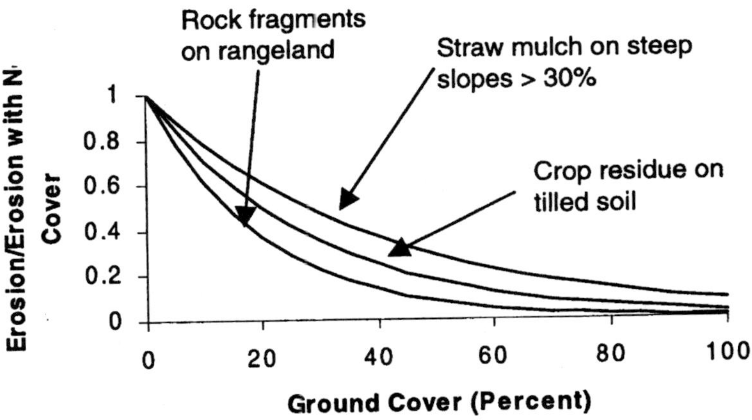

Figure 2.7 Effect of ground cover on erosion by raindrop impact and surface runoff.

tion site (Box, 1981; Meyer et al., 1972). Rock cover on rangelands can significantly increase infiltration as rock cover increases (Simanton et al., 1984). Straw mulch applied to construction sites is less effective in reducing erosion than crop residue in a farm field. Crop residue from last year's crop is critically important in conservation tillage systems used widely to control erosion on cropland (Chapter 7).

Cryptogams are collections of mosses, algae, lichens, and liverworts that develop on soil surfaces and in soil profiles. Organic soil crusts resulting from cryptogams can reduce infiltration rates by blocking the flow through the macropores of a soil, increasing runoff and soil erosion. In other cases, cryptogams can enhance porosity and infiltration by increasing soil aggregation and roughness (e.g., West, 1990). Cryptograms also protect the soil from erosion by direct raindrop impact and the erosive forces of surface runoff.

Mechanical Soil Disturbance

Soil management, in addition to the inherent soil properties greatly affects soil erodibility. The presence of organic material, discussed earlier, is one way that management affects erodibility. Mechanical disturbance also affects both soil erodibility and the erosivity of raindrops and runoff. Mechanical disturbance that creates surface roughness slows runoff and reduces its erosivity. Surface roughness creates depressions that pond water, reduce the erosivity of raindrops, and store locally deposited sediment. Surface roughness can create erosion-resistant clods that reduce erosion and increase infiltration.

Mechanical soil disturbance can also increase soil erodibility. For example, soil in a seedbed condition is about twice as erodible immediately after tillage than it is over an extended period of a few years as the soil consolidates. Mechanical soil disturbance also buries surface plant materials and mixes those materials with belowground plant residue to greatly reduce erosion.

Soil can be formed into ridges that redirect runoff from a downslope path to a path around the slope, a practice known as *contouring* (Chapter 7). The redirection reduces the erosivity of the runoff and thus erosion by flow. If the grade along the furrow is sufficiently flat, some of the sediment eroded by raindrop impact on ridge sideslopes can be deposited in the furrows, which also reduces sediment yield from the hillslope (Foster et al., 1997; Meyer and Harmon, 1985). The effectiveness is greatest with high ridges perfectly on the contour and where high rainfall intensities do not occur to cause overtopping of the ridges (Moldenhauer and Wischmeier, 1960).

Terrace channels are placed on hillslopes to reduce slope lengths to control erosion by surface runoff. Runoff flows down the interterrace area to be collected by the terrace channel and diverted around the slope at a nonerosive velocity. Deposition occurs in the terrace channel that causes the land to become benched over time in a stair-step configuration (Foster and Highfill, 1983).

Terraces can also be created as a part of an operation known as *chaining* on rangelands to remove trees and shrubs. A large-link chain or cable is stretched between two track vehicles driving in parallel across the ground surface and ripping the taller vegetation from the soil. The resulting brush can be pushed into windrows approximately on the contour. These windrows function as terraces when they intercept and direct surface runoff around the hillslope.

WIND EROSION

Wind erosion occurs when the forces applied to the soil by wind are greater than the resistance of the soil to these forces. The forces are directly a function of the environmental conditions at a particular location where wind erosion is occurring. Wind erosion is a function of the amount of sediment produced by detachment processes and the transport capacity of the wind (Chapter 4). The four factors of climate, soil, topography, and land use determine wind erosion at a site. Wind erodes soil particles by saltation, suspension, and creep along the soil surface (Figure 2.8).

Arid lands comprise about one-third of the world's total land area and are the home of one-sixth of the world population (Skidmore, 1994). As noted in Chapter 1, the areas most susceptible to wind erosion are agricultural regions of North Africa and the Near East, parts of southern and eastern Asia, the Siberian Plains, Australia, southern South America, and portions of the United States.

The Great Plains of the United States and the Prairie Provinces of Canada are areas of North America where conditions conducive to wind erosion are prevalent. Figure 2.9 shows the areas of significant wind erosion in the continental United States. Few regions of the United States or elsewhere, however, are safe from wind erosion when the protective vegetation cover is disturbed, as is the case at mining or construction sites. The sod busting (disturbance of the continuous grass cover due to land-

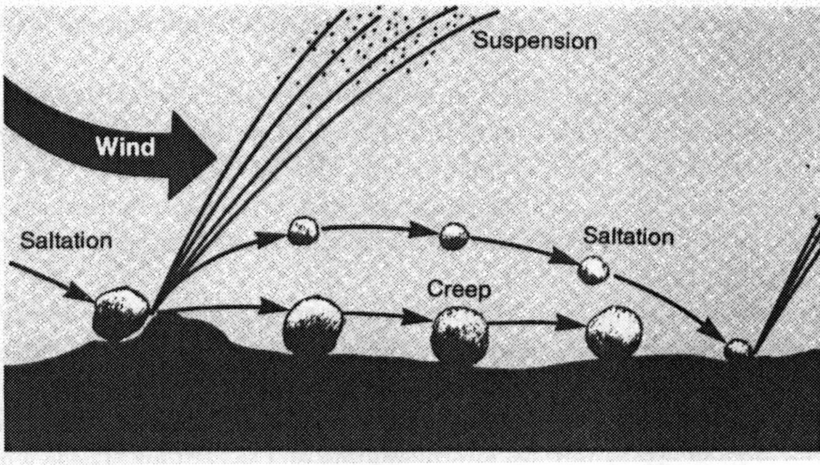

Figure 2.8 Wind erosion transport modes of creep, saltation, and suspension. (From USDA, SCS, 1989.)

use conversion to agriculture) in Oklahoma and adjacent states during the 1930s led to accelerated wind-erosion rates and the American Dust Bowl, with extraordinary dust storms that transported airborne sediment particles more than 2000 miles (3300 km). Although this land degradation and air pollution contributed to the economic difficulties of the Great Depression in the United States, it also served as one of the catalysts for the

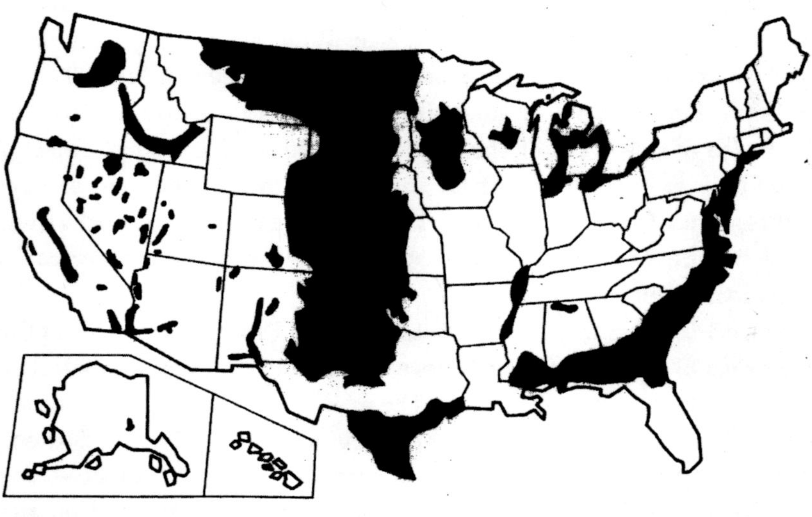

Figure 2.9 Wind erosion problem areas in the continental United States. (From USDA, SCS, 1989.)

establishment of soil-erosion research and soil-conservation programs at the federal level of government. These new programs eventually resulted in major changes in farming practices throughout much of the United States. Eroding land in the Great Plains was estimated to contribute between 250 million and 75 million tons (220 million and 70 million metric tons) of dust per year to the atmosphere in the 1950s and 1960s, respectively.

Climate

The major climatic variable affecting wind erosion is wind velocity and the duration of a wind storm. The wind-erosion rate is a direct function of wind velocity and multiplied by the duration of an event, provides an estimate of the amount of wind erosion for a single storm. The wind-erosion rate is approximately proportional to the wind velocity cubed. The erosivity of wind can be computed from measured wind data. Wind erosivity varies by location, because wind is more prevalent in some locations than in others. For example, the erosivity of wind at Grand Forks, North Dakota, is 1.4 times that at Goodland, Kansas. Wind erosivity at Jackson, Mississippi is about 15 percent of that at Goodland, Kansas. Thus, if all other conditions are the same between locations, erosion would be directly proportional to the wind erosivity.

Wind erosivity also varies during the year. For example, peak erosivity occurs in late winter and early spring and is lowest in late summer at many locations. Wind erosion is greatest when peak erosivity occurs at the same time that the land is most vulnerable to wind erosion, which is usually when vegetative cover and plant residue cover are minimal and soil aggregates are fine, loose, and dry. Wind erosion is reduced by providing protection during the peak erosive period. Wind erosivity also varies from year to year. Although the within-year erosivity is sufficiently repeatable that it can be considered in selecting control measures, the year-to-year variations are not predictable and cannot be planned for.

Wind direction varies from storm to storm and by season. Wind directions and the relative magnitude of wind erosivity for each direction and the probabilities of the directions have been determined from weather data and are available for use in conservation planning (Skidmore, 1987; Skidmore et al.,1994; USDA, NRCS, 2001b). The reason that wind direction is important is that the orientation of wind with ridges and vegetation in rows and strips greatly affect erosion. Vegetation has a much greater effect on reducing wind erosion when the ridges and vegetation rows are oriented perpendicular to the wind direction (Skidmore et al.,1994). Wind direction is important with respect to the field's long and short dimensions. Erosion is increased when the wind direction is parallel to the field's long dimension.

Other climatic variables, especially temperature and precipitation, have both primary and secondary effects on wind erosion. These variables affect vegetation growth and decomposition of plant residues. A primary effect is that soil moisture immediately at the soil surface is critically important to wind erosion, because the presence of soil moisture can reduce the soil's erodibility from wind (Skidmore et al., 1994). Sometimes, wind velocity, precipitation, and temperature are combined to form a climate erosivity index for wind erosion. Precipitation, wind velocity, solar radiation, temperature, and other climatic factors that affect soil moisture at the soil surface greatly affect wind erosion. Wind erosivity can be very high in a particular location, but wind erosion is not a problem because precipitation is sufficient to keep the soil moist, especially during periods of high erosivity. The probabililty distribution of wind erosivity between years is only a part of what determines wind erosion probability from year to year. Even though annual rainfall is high in a particular region such as along the Atlantic coast in the southeastern United States, dry periods usually occur each year, and with high temperature and easily erodible sandy soils, wind erosion can be a problem in locations that would not be expected to have wind erosion based on average annual conditions.

Soil

Soils vary in their inherent susceptibility to wind erosion as a function of intrinsic properties. A standard condition is established to measure wind erosion and the relationship by soil-texture classes (Appendix A). An estimate of the percent of dry-soil aggregates greater than 0.03 in. (0.8 mm) determined by a standard test procedure is a measure of soil erodibility. Erodibility is closely related to the soil aggregate sizes because aggregate size is a measure of the force required to transport soil particles. The size of aggregates produced by this test is also a measure of soil strength and its resistance to the forces applied by wind and by windblown sediment that detaches soil particles. Each soil-particle size and density has a threshold shear stress. Wind erosion does not occur unless shear stresses from the wind exceed this threshold. The threshold shear stress increases as particle size and density increase. This relationship is valid for both primary particles that have a somewhat greater density than the density of soil aggregates.

Soil erodibility varies by season because of changing conditions within the year. For example, some times of the year are typically drier than are other times. Soil erodibility is greater during dry periods because increased soil moisture greatly reduces soil erodibility. Other factors, such as biological activity in the soil related to soil moisture and temperature, affect soil erodibility. Exposed frozen soil that freeze-dries during the winter in the north-central United States leaves soils highly erodible in late winter and early spring. The variation in erodibility and the climate variable effect on temporal soil variability depends on soil texture. In general, erodibility increases from fall to spring, but the greatest increase occurs in fine-textured soils. Clay is the least erodible of the soil textures in the fall but is about as erodible as a sandy loam, a soil considered to be highly erodible in spring. Sandy loam soils are typically highly erodible in both spring and fall and do not vary nearly as much in erodibility during the year as clay soils. Intermediate-texture soils are intermediate in relative erodibility in both spring and fall (Skidmore, 1994).

Soil erodibility can vary significantly from year to year. One of the reasons for the severe wind erosion during the Dust Bowl days (1930s) was an unusual combination of high wind erosivity with a drought that made the soils more erodible than would ordinarily be the case. On top of this already highly erodible situation, the drought had greatly reduced vegetative cover that ordinarily would be available to protect against the wind's erosive forces. Given the soil texture of the Great Plains region, the situation was perfect for a wind-erosion disaster.

An important climate and soil interaction is *crusting* (Sumner and Stewart, 1992), which also interacts with land management. Crusting is a dense layer at the soil surface that can reduce wind erosion dramatically on fine-textured soils. Although crusts are formed in several ways, one primary way that soil crusts form is by raindrop impact. Thus, many of the variables important in water erosion, including kinetic energy, rainfall intensity, and rainfall amount, are important in crust formation, and hence wind erosion.

Topography

Topography also affects wind erosion. One important topographic variable is field length. Up to a point, wind erosion increases as field length increases. Thus, one way to reduce wind erosion is to reduce field length. The other effect of topography on wind erosion is caused by hilltops that protrude into the wind and obstructing airflow. A hilltop affects the erosive forces of wind along the flow path. On the windward side of a hilltop, the forces acting on the surface increase as the wind is forced upward by the hill. This increase in forces increases wind erosion on the windward side of the hill. On the leeward hillside, the forces acting on the soil are reduced, which can result in deposition, especially for soil easily eroded, as with sand dunes.

Soil properties vary over the landscape, which results in a variation of erosion because of spatial soil erodibility changes. Soil moisture and temperature vary by their position on the landscape, which affect vegetation growth and other factors that affect wind erosion. Wind erosion varies over the landscape in a corresponding way. Wind erodibility by soil texture class is shown in Table 2.1. Also shown in this table is an estimate of the percent of dry-soil aggregates greater than 0.03 in. (0.84 mm) and an estimate of the wind erodibility index. This table is used to estimate the amount of wind erosion for a specific location when estimates of wind erosivity and site characteristics are known.

Land Use

Vegetative cover, surface roughness, rock cover, and crusting are major factors that influence wind erosion. Land-use activities that affect these factors have major effects on erosion.

Vegetation

As with water erosion, vegetation reduces wind erosion by reducing the forces that the wind applies to the soil surface (reduced erosivity) and by increasing the resistance of the soil to erosion (reduced erodibility). Standing vegetation, including live growing vegetation and standing stubble from harvested crops, has a major effect on reducing wind erosion. Even low amounts of vegetation can reduce wind erosivity significantly. The practices and situations that produce high amounts of tall, dense biomass with little open space are most effective at reducing wind erosion. Vege-

Table 2.1 Descriptions of Wind Erodibility Groups

Wind-Erosion Group	*Predominant Soil Texture Class of Surface Layer*	*Dry-Soil Aggregates > 0.84 mm (%)*	*Wind Erodibility Index (Mg/ha)*
1	Very fine sand, fine sand, or coarse sand	1	695
		2	560
		3	493
		5	404
		7	359
2	Loamy very fine sand, loamy fine sand, or sapric soil material	10	300
3	Very fine sand loam, fine sandy loam, sandy or coarse loam	25	193
4	Clay, silty clay, noncalcareous clay loam, or silty loam >35% clay content	25	193
4L	Calcareous loam and silt or clay loam and silty clay loam	25	193
5	Noncalcareous loam and silt loam with <20% clay or sandy loam	40	126
6	Noncalcareous loam and silt loam >20% clay	45	108
7	Silt, noncalcareous silty clay loam <35% clay and fibric organic	50	85
8	Soils not susceptible to wind	>80	0

Source: Data from Skidmore (1994).

tation varies in its effectiveness for controlling erosion during the year. The erosion control objective is to have maximum vegetative cover during periods of the year when wind erosivity is greatest. Roots in the soil help hold the soil in place mechanically. Root decay increases organic matter, which increases soil resistance to wind erosion.

Strips of tall, dense vegetation placed perpendicular to the dominant wind direction greatly reduce wind erosivity and erosion. Barriers of trees, shrubs, tall-growing crops, and grasses placed around field boundaries reduce wind erosivity at the barrier and for a considerable distance beyond the barrier (Hagen, 1976; Skidmore and Hagen, 1977). A disadvantage of barrier systems is that they occupy space that otherwise could be used for crop production. Tree-barrier systems themselves can be designed to provide valuable crops, such as nuts, fruits, or wood. Perennial barriers that grow slowly often are difficult to establish (Skidmore, 1994). These barriers also compete for water and plant nutrients.

Ground Cover

Stable material, including crop residue and rock lying on the soil surface is highly effective at reducing wind erosion. This material reduces the erosive forces applied to the soil, which reduces wind erosion. The material also reduces evaporation, which increases soil moisture at the soil surface and in turn reduces soil erodibility. Surface cover can be naturally occurring, such as rock on rangeland or rock in cropland soils that is brought to the surface with tillage. Straw mulch and manufactured materials can be applied to construction sites to control wind erosion. Another important source of ground cover on range and pasture lands is the litter produced by senescence from permanent vegetation. Perhaps the most important source of ground cover is the crop residue left on the soil surface after crop harvest. The amount of organic materials on the soil surface and placed in the soil is a function of decomposition, which varies by location because of temperature, precipitation, and soil differences.

In many arid and semiarid regions, rock covers protect the soil surface on rangelands and similar lands to greatly reduce wind erosion. These covers can be produced by various processes (Toy and Osterkamp, 1999), such as the selective removal of fine-textured particles from the soil surface by both water and wind. Sometimes, the surface is "hardened" by a layer of close-fitting fragments and pebble to produce a *desert pavement*.

Chemicals known as *soil stabilizers* (e.g., polysaccharides) are cost-effective for applying to small areas such as construction sites to control wind erosion (Armbrust and Lyles, 1975; Lyles et al., 1969). Ideally, these soil stabilizers provide erosion protection while allowing water infiltration to increase soil moisture and permit vegetation growth. Freeze/thaw or wet/drying cycles tend to reduce the effectiveness of soil stabilizers for erosion control. Often, water alone is spread on soil surfaces or mine haul roads for temporary wind-erosion control.

Mechanical Soil Disturbance

Mechanical soil disturbance can be both a blessing and a curse to wind erosion control. The mechanical disturbance usually done by tillage can create a rough soil with both a high degree of random roughness and large stable clods that are resistant to erosion. Over time, rainfall decays the roughness and the depressions fill with sediment. The roughness must be re-formed to regain erosion control lost with roughness decay. Conversely, tillage that breaks up a crust and leaves fine soil particles on the surface greatly increases erosion. Crusts can be very resistant to wind erosion, but a mechanical soil disturbance destroys the crust and leaves the soil highly susceptible to erosion.

Tillage can also be used to create ridges that reduce wind erosion significantly when the ridges are oriented perpendicular to the predominant wind direction. Ridges parallel with the wind direction are ineffective at controlling wind erosion. Over time, ridges are eroded and the furrows between ridges fill with sediment so that the ridges lose their effectiveness.

INTEGRATED SITE PERSPECTIVE

In this chapter we have chosen the context that both wind and water erosion are functions of climate, soil, topography, and land use. The influence of each factor can be described as if they act independently, when in actuality, the factors interact with each other. In simplest terms, erosion is a function of erosivity in relation to erodibility, but each of the factors contains elements of both erosivity and erodibility. For example, erosivity is determined by climate, but climate also affects erodibility by affecting soil moisture and soil crusting. Soil erodibility can be measured under standard conditions, but the actual erodibility at the time of the effect is different because of soil moisture and the presence of organic material in the soil from plant roots. Another perspective in considering erosion is that it is a function of the forces applied to the soil in relation to the forces within the soil resisting the erosive forces. Erosion can be reduced in one of two ways: to reduce the force being applied to the soil, or to increase the soil's resistance to erosion. The fundamental question is always: How do each of these factors affect the forces that control erosion for the site in question?

Another important question at each site where both wind and water erosion are significant is how the two interact. These two types of erosion occur at different times: wind erosion when the soil is dry and water erosion when the soil is bare. Should the two types of erosion be added to determine a total erosion for the site? Erosion is frequently expressed as an average soil loss, which is usually a field length for wind erosion and a slope length for water erosion. In some fields, the sediment blown from one field is deposited on a nearby field. Some of this sediment can be eroded by a subsequent storm and deposited back on the original field. This multiple erosion and deposition may not greatly reduce soil depth, but these multiple events can be highly selective and enrich the soil in coarse particles, which degrade the soil. Whether or not the erosion from both wind and water erosion on the same site is added to determine overall erosion depends on the specific site and how wind and water erosion occur with respect to each other over the area.

SUMMARY

The factors affecting water and wind erosion are climate, soil, topography, and land cover and use. These factors define the environmental settings in which erosion processes operate. Human activities associated with agriculture, mining, and with residential, commercial, and highway construction remove the protective vegetative cover, resulting in accelerated erosion by both water and wind. Establishment and maintenance of vegetative covers and plant residues are central to effective erosion control in most cases.

Scientists and engineers have investigated the factors affecting erosion for many years. Most of the early work focused on erosion resulting from average annual, seasonal, or monthly climate conditions, or perhaps, the erosion caused by an individual storm. The study of dynamic soil-erosion processes by water and wind has increased dramatically in recent years. Current computer technology allows detailed examination of complex physical relationships and integration of the numerous variables that are responsible for the temporal and spatial variabilities of erosion processes, as discussed in Chapters 4 and 5. New developments have also become available for wind-erosion control, as discussed in Chapter 7.

SUGGESTED READINGS

Commission on Long-Range Soil and Water Conservation, Board on Agriculture, National Research Council. 1993. *Soil and Water Quality: An Agenda for Agriculture*. National Academy Press, Washington, DC.

Commission on Watershed Management, National Research Council. 1999. *New Strategies for America's Watersheds*. National Academy Press, Washington, DC.

El-Swaify, S. A., W. C. Moldenhauer, and A. Lo. 1983. *Soil Erosion and Conservation*. Soil and Water Conservation Society, Ankeny, IA.

Renard, K. G., G. R. Foster, G. A. Weesies, D. K. McCool, and D. C. Yoder. 1997. *Predicting Soil Erosion by Water: A Guide to Conservation Planning with the Revised Universal Soil Loss Equation (RUSLE)*. USDA Agricultural Handbook 703. U.S. Government Printing Office, Washington, DC.

3

Types of Erosion

The major classification of erosion type is by erosive agent, wind or water, that causes the erosion. Other agents, such as gravity and tillage, cause soil movement and are mentioned briefly in this book. The most important type of water erosion is that caused by rainfall, surface runoff from rainfall, and surface runoff from irrigation. Subtypes of water erosion are classified based on spatial context and topographic position within a watershed. Subtypes of wind erosion are defined by mode of sediment transport. The two primary modes of sediment transport by wind are (1) creep plus saltation, which transports medium-sized and coarse sediment; and (2) suspension, which transports fine sediment.

WATER EROSION

Spatial Context

Water erosion is best examined within the spatial context of a watershed (drainage basin, catchment). Watersheds range in size from less than an acre (hectare) to thousands of acres (hectares), including large river basins. The smallest and simplest watershed is composed of overland flow areas adjacent to a single channel (Figure 3.1.). A large watershed consists of a "nested hierarchy" of smaller watersheds linked by a concentrated flow (channel) network (Figure 3.1).

Water flow and its paths are central to the study of water erosion. Water flows in two types of conduits, open channels and pipes (Figure 3.2). An open channel has a free-water surface exposed to atmospheric pressure, while pipe flow fills the conduit with water that flows under hydraulic pressure (Chow, 1959). Open-channel flow occurs in rills, gullies, and

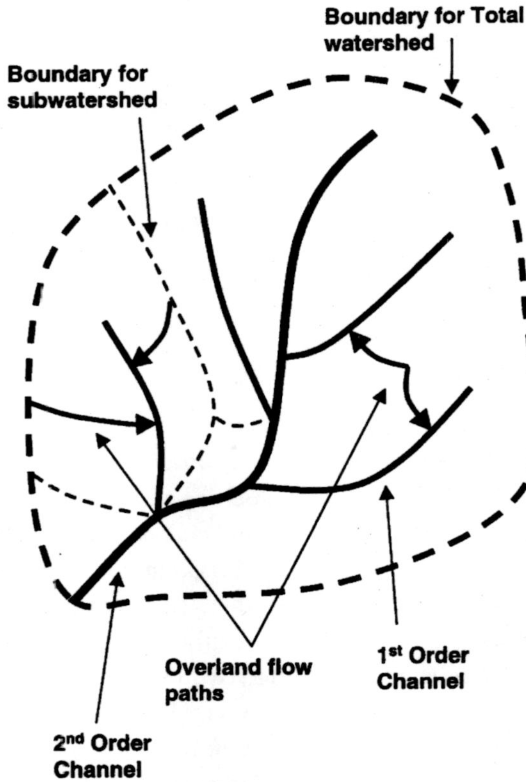

Figure 3.1 Simple and nested hierarchy of watersheds.

stream channels, for example. Pipe flow occurs through the soil macropores in saturated soil.

The cross-sectional shape of open channels varies greatly (Figure 3.3). These channels range from deeply incised channels, such as permanent gullies, to broad, shallow channels, such as ephemeral gullies. One mea-

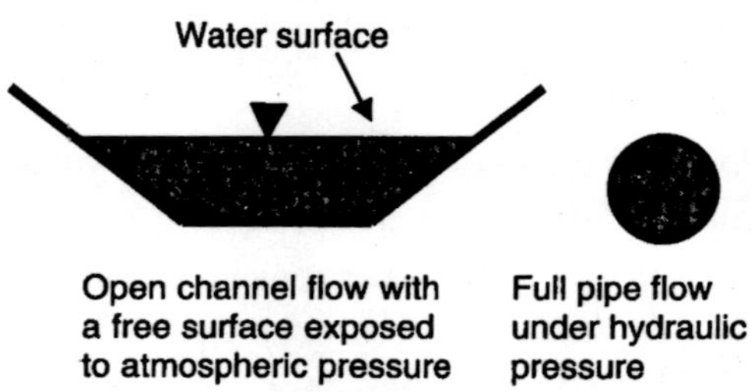

Figure 3.2 Open channel and pipe flow.

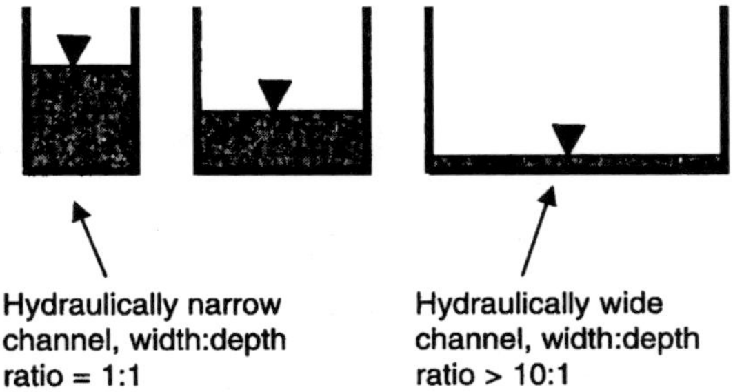

Figure 3.3 Channel width/depth ratios.

sure of channel shape is the width-to-depth ratio. For a narrow channel this ratio is small (e.g., 1 : 1), whereas for a wide channel it exceeds 10 : 1 and for sheet flow it is infinite.

Channels develop as a natural part of landscape evolution. Areas between a channel and the divides that delineate the channel's drainage area are the overland-flow portions of the landscape, illustrated in Figures 3.1 and 3.4. Hydrologists often consider runoff on overland-flow areas to be sheet flow, although sheet flow occurs only on smooth surfaces. The

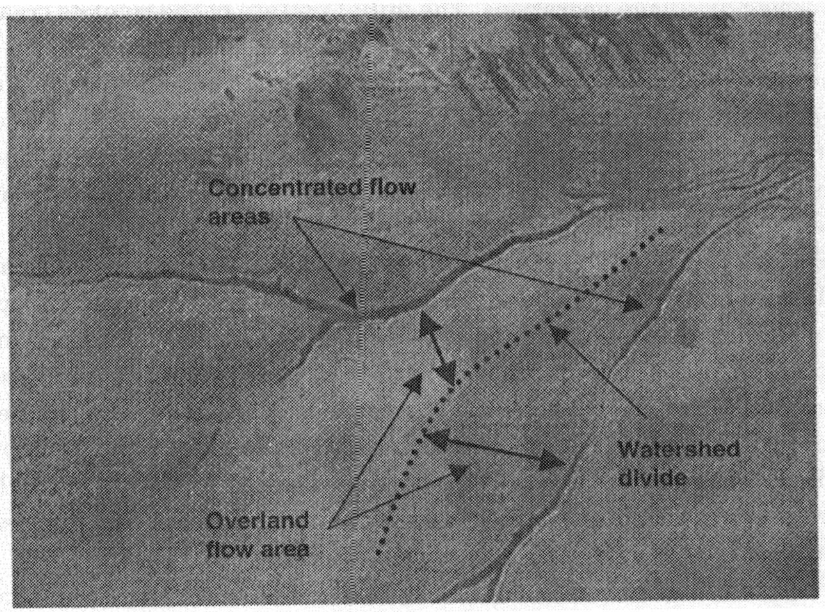

Figure 3.4 Overland and concentrated flow areas on a landscape. (Courtesy of USDA, NRCS.)

depth of runoff over most natural and mechanically disturbed soil surfaces varies greatly with location on a hillslope.

Interrill–Rill Areas

Although overland flow typically is visualized as being a broad sheet flow, runoff actually is concentrated in many small rivulets of water (Emmett, 1970; Foster 1971). These small rivulets are referred to as *rill areas* even if no erosion takes place in them (Foster, 1971; Meyer et al., 1975b). If any erosion by flow occurs, it will occur in these areas, and this erosion is defined as rill erosion. The areas between the rill areas are the *interrill areas*, and the erosion that occurs on these areas is defined as interrill erosion. The interrill and rill areas together make up the overland-flow areas of landscapes. Surface runoff, known as *Hortonian flow*, on these areas is produced when rainfall intensity exceeds the infiltration capacity of the soil (Appendix B; Horton, 1933). Interrill plus rill erosion is the total water erosion that occurs on the overland flow areas of the landscape.

The location of the rill areas and their pattern are determined by the microtopography of the soil surface on the hillslope, not by the macrotopography of the landscape. On cultivated hillslopes, the microtopography is formed by tillage operations. The initial pattern of the rivulets created by tillage can evolve into a network of rills and small channels. By the traditional definition, rills are channels that are so small that they can be obliterated by normal tillage operations. After obliteration, rills tend to form in new locations. The surface of a mechanically graded hillslope can appear uniform, but intensive erosion by runoff on these hillslopes often produces rills (Figure 3.5).

Surface runoff also occurs in rivulets of water on natural, noncultivated hillslopes even when well-defined incised rills do not form. The flow pattern often is determined by plant stems and roots, debris, rocks, and local deposition that create an uneven surface so that runoff is concentrated into small channels between the obstructions. Vegetation also can be so dense that runoff is nearly continuous across the hillslope with characteristics of sheet flow. Overall, the concept of runoff occurring in rivulets of water on overland-flow areas is a very powerful concept that applies in most cases, especially where significant erosion by flow occurs.

Surface runoff and sediment flow laterally on interrill areas to adjacent rill areas. The slope lengths of interrill areas often are short, less than 3

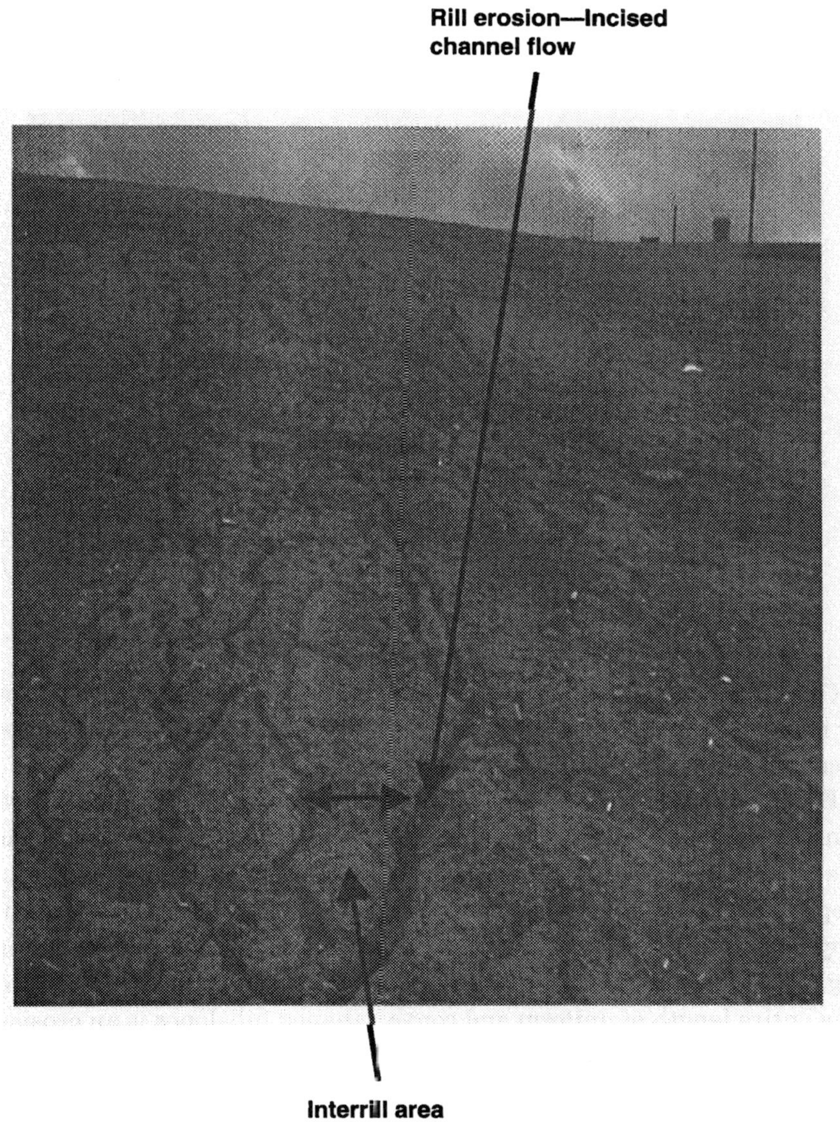

Figure 3.5 Highly rilled construction site.

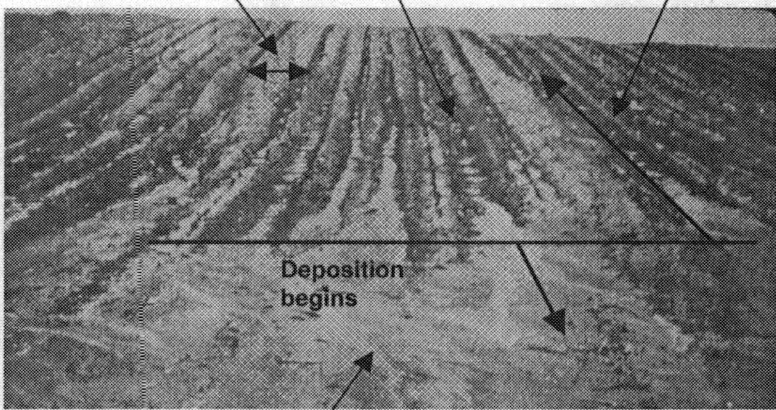

Figure 3.6 Erosional and depositional areas on a hillslope.

ft (1 m). Interrill areas are delineated so that all detachment on them is by raindrop impact. Similarly, rill areas are delineated so that all detachment in them is by surface runoff.

Erosional and depositional regions can be identified for hillslopes with concave segments that experience deposition (Figure 3.6). Net erosion, specifically defined as soil loss, occurs on the upslope portion of the hillslope, while net deposition occurs on the downslope portion of the hillslope. The amount of sediment that leaves a hillslope with a depositional region is less than the soil loss from the erosional region of that hillslope. The entire length of uniform and convex-shaped hillslopes is an erosional region, such that the amount of sediment leaving these hillslopes equals the soil loss. Also, hillslopes with only slight concavity do not experience net deposition, and thus the amount of sediment leaving such hillslopes also equals the soil loss. Deposition can be induced by means other than hillslope shape. Dense vegetation and other barriers that slow runoff dramatically can induce deposition on any slope shape.

Sheet and Rill Erosion

Sheet and rill erosion also are types of water erosion that occur on overland-flow areas. Although this older erosion classification often is used interchangeably with interrill and rill erosion, the two classification systems are conceptually different. The interrill–rill classification is based on a spatial context that defines interrill and rill areas on hillslopes. The sheet and rill erosion classification is based on a concept of progressive erosion severity. Sheet erosion, which is a uniform removal of soil from the surface, is assumed to be the first phase of the erosion process, and sheet-erosion rates are assumed to be low. As erosion becomes increasingly severe, rill erosion is assumed to begin. Rill erosion progresses to gully erosion, which produces deeply incised channels.

A rule of thumb is that rill erosion begins when sheet erosion reaches about 6 to 7 tons/acre (15 metric tons/ha) per year. Interrill erosion actually is sheet erosion because it is uniform over the interrill area. This erosion rate can reach 20 tons/acre (40 metric tons/ha) per year on ridges separated by furrows without rill erosion occurring in the furrows, indicating that rill erosion is not necessarily an advanced form of interrill erosion (Meyer and Harmon, 1985). In fact, deposition can occur in the furrows if the grade along the furrows is relatively flat when interrill erosion is high.

Concentrated-Flow Areas

A *concentrated-flow area* is not distinguished from a rill area on the basis of channel size but by its position within the landscape, and the role of macrotopography in determining the location of the concentrated flow channel illustrated in Figures 3.1, 3.4, and 3.7. The location of a concentrated-flow channel is determined solely by the macrotopography of the landscape, and if the channel is filled by tillage or earth-moving equipment, the concentrated-flow channel is formed again in its previous location. Although the location of natural concentrated-flow channels is determined by macrotopography, constructed concentrated-flow channels can be placed on the hillside to reduce slope length and, in turn, rill erosion. Examples include terrace channels on farm fields and diversion ditches on construction sites.

Ephemeral Gully Erosion

Ephemeral gullies generally occur within field-sized areas where farming and similar land-disturbing operations take place. The amount of sediment produced by this type of erosion can equal the amount of sediment

Figure 3.7 Blending of ephemeral gully areas with overland flow areas. (Courtesy of USDA, NRCS.)

produced by interrill and rill erosion in the same field (Foster, 1985; Thomas et al., 1986). In farm fields, these channels are crossed as a part of routine farming operations and are filled routinely by tillage operations that move soil from the overland-flow areas adjacent to the ephemeral gullies. The macrotopography of the surface causes ephemeral gullies to form again in the same location after refilling by farming operations. This periodic refilling and reformation by erosion is the reason for the name *ephemeral gully*. Through time, these channels gradually become blended with the hillslopes rather than remaining incised with vertical sidewalls (Figure 3.7).

Loosening of soil by tillage leaves a surface soil layer in the ephemeral gully area that is much more erodible than the untilled soil immediately beneath the tilled zone. Flow quickly erodes the ephemeral gully to the depth of the nonerodible, untilled layer, and then the flow erodes the sidewalls of the gully, producing a wide, shallow channel with a high width-to-depth ratio.

Although ephemeral gully erosion is a feature unique to cultivated fields, a similar type of erosion can occur on reclamation and construction sites, where a loose soil layer is prepared for a seedbed but immediately beneath the surface layer is compact soil caused by vehicular traffic that is resistant to erosion. Flow in concentrated-flow areas erodes rapidly through the surface soil layer and reaches the resistant soil beneath. Erosion widens rather than deepens the channel. Ripping is used to break up the underlying compacted soil.

Permanent, Incised Gully Erosion

Permanent, incised gullies occur on both natural and disturbed lands, (Figure 3.8). Permanent, incised gullies in agricultural fields are defined as channels that are too deep to cross or to fill with normal farming operations (Foster, 1985). Land-use change that increases runoff significantly or creates a lowered base level with an overfall of water from a lateral channel often are responsible for gully development. Permanent, incised gullies commonly are recent in age, developing in just a few years. Permanent gullies typically are incised channels that are wide and deep relative to the flow in them (Heede, 1975; Piest et al., 1975). Upstream advance of a headcut creates a permanent, incised gully. The advance of

Figure 3.8 Permanent, incised gully in an agricultural field. (Courtesy of USDA, NRCS.)

a gully into a field causes severe damage to the field and produces high sediment loads.

Not only does the channel bottom drop in elevation abruptly at the headcut, the channel may also abruptly widen from the upstream concentrated-flow area to the downstream permanent, incised gully, (Figure 3.9). These gullies typically have near-vertical sidewalls (banks) when they form in cohesive materials. The sidewalls also retreat laterally to produce a wide channel with steep sidewalls and shallow flow.

Erosion in permanent, incised gullies is episodic, varying from year to year (Piest et al., 1975). For example, vegetation can develop on the slumped material in a gully, protecting this material from removal by flow. The weight of the slumped soil stabilizes the base of the sidewall, preventing further slumping into the gully. High flow from an infrequent but high-magnitude runoff event can breach the protection provided by the vegetation at the base of the gully sidewall, remove the slumped soil, and destabilize the gully sidewall once again. The gully now is susceptible to erosion by much smaller runoff events than would have been the case had the highly erosive, destabilizing runoff event not occurred.

Stream-Channel Erosion

Stream channels are an integral part of the landscape. Stream channels develop in the absence of human activities, but activities on upland areas and within channels themselves can greatly influence stream-channel erosion. Channel features, including grade and meander form, adjust to accommodate the flow and sediment load delivered to the channels from the upland areas (Schumm, 1977). Therefore, changes in land use that modify runoff and sediment delivery produce changes in the stream channels. Stream channels are changing constantly, but the change in stable stream channels can be almost imperceptibly slow (Trimble, 1977).

Abrupt changes in land use, such as forestry to urban use or forestry to intensive agriculture, that increase upland runoff significantly may destabilize stream channels and initiate channel erosion. This erosion widens the channels, and headcuts may form that migrate upstream rapidly, producing large sediment loads and degrading the stream quality seriously. The most active locations of stream-channel erosion usually is on the outside of meander bends, where the channel bank can retreat many feet (several meters) during severe storms. Channel erosion can be controlled by reduction of runoff rates with impoundments, construction of

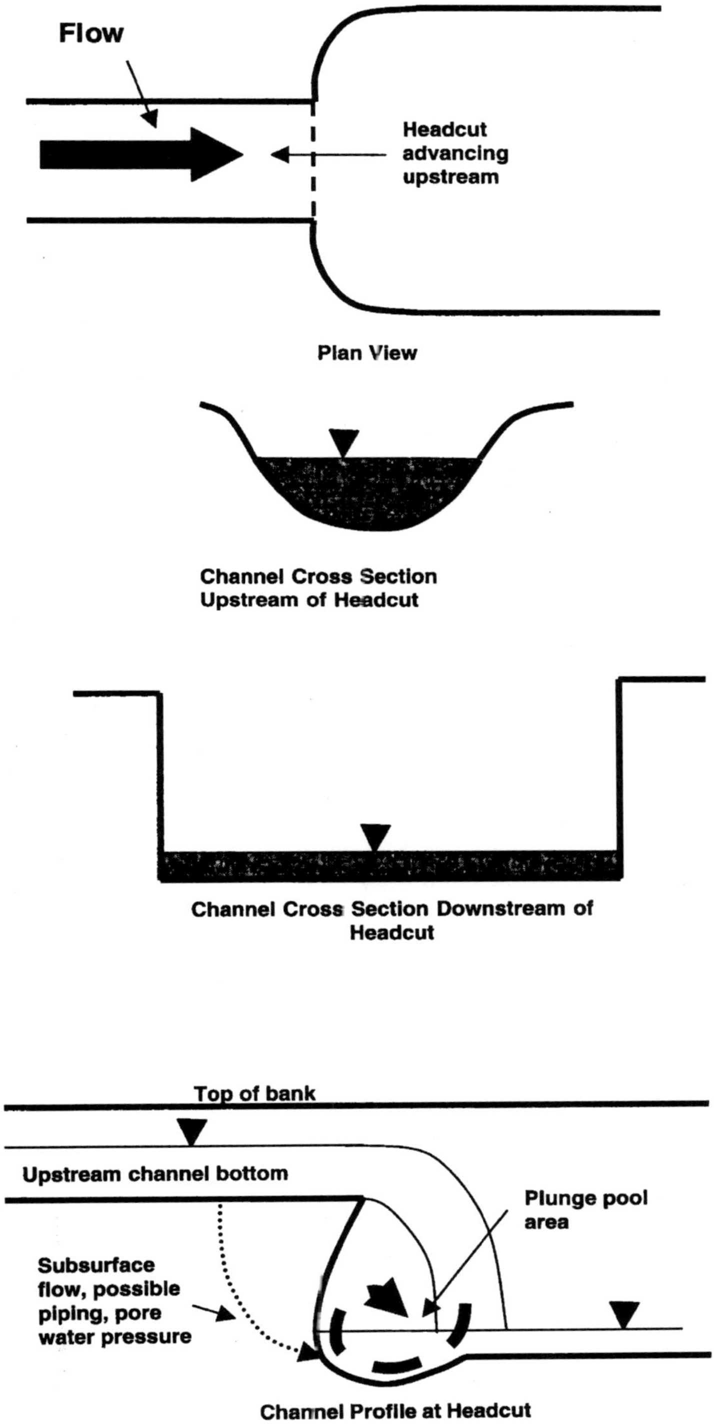

Figure 3.9 Effect of headcut on gully geometry.

enlarged channel cross sections, installation of grade-control structures in the channel, addition of bank protection, and placement of in-stream vanes to divert flow away from channel banks (Chapter 7; Shields et al., 1995).

Stream channels in undisturbed forestlands typically are stable, but logging and road building can disturb these streams drastically (Dunne, 1998; Elliot, 1999; Grace, 2000; Warrington et al., 1980). The land disturbance increases surface runoff, stream discharge rates, and the sediment loads delivered to the stream channel. Drainage ditches built along roadsides often convey the increased runoff and sediment produced by erosion on roadways, skid trails, landings, and other disturbed areas directly to stream channels. Undisturbed buffer zones along a stream can be retained and drainageways can be constructed to direct runoff through these areas to filter the sediment in order to protect a stream.

Stream-channel instability can also be caused by dramatic increases in the sediment load reaching streams from upland areas. Agricultural development in the nineteenth and early twentieth centuries exposed large land areas in select parts of the United States, including the Piedmont region in the southeastern United States and the Coon Creek Watershed in western Wisconsin (Trimble, 1977). In both regions, high rates of interrill, rill, and gully erosion on newly cleared and cultivated lands produced and delivered large amounts of sediment to valley streams, where much deposition and aggradation occurred.

In the early 1930s, the USDA conducted projects in several areas to demonstrate erosion-control technology. Vast cropland areas in the southeastern United States once in forest were converted back to forestland, and extensive conservation farming practices were applied in the Coon Creek watershed, which dramatically reduced upland erosion and sediment delivery to stream channels. The stream channels, which had first adjusted to the increased sediment loads from the high rates of upland erosion by storing sediment, adjusted once again to the new reduced rates of upland erosion by eroding previously stored sediment (Trimble, 1977, 1983).

Watershed Sediment Yield

Watershed sediment yield is the amount of sediment delivered at the outlet of a watershed. Sediment yield estimates are used to design reservoirs and to analyze sedimentation and water-quality problems. Sediment yield is a measure of average net erosion for the watershed. Sediment yield is the sum of the sediment produced by all erosional sources, including that from overland flow, ephemeral gully, permanent, incised gully, and stream channel areas less the amount of sediment deposited on these areas and on the valley floodplains.

Erosion by Irrigation

Irrigation water is applied to the land by overhead sprinklers, surface-applied flow, or subsurface emitters. When irrigation water is applied properly by sprinkler or subsurface systems, erosion is not a problem. Erosion can be serious, however, with surface-applied irrigation water, an irrigation method widely used in the western United States. With this type of irrigation, surface water is introduced at the head of furrows to flow down the field and infiltrate into the soil. Erosion rates with this irrigation system have been large, especially at the upper ends of the furrows, where erosion rate can be four times the average erosion rate for the field (Trout, 1996).

Erosion by Snowmelt and Rainfall on Thawing Soil

Erosion by overland flow resulting from snowmelt can occur, but this erosion depends on a particular combination of conditions. If snowmelt runoff happens to occur when the soil is thawing, erosion can be substantial (McCool et al., 1995; 1997; Van Klaveren and McCool, 1998). A much more common occurrence is rainfall on thawing soil, which routinely happens in an area in the northwestern United States known as the Palouse Region or the Northwest Wheat and Range Region (Austin, 1981). The soil is highly erodible, and even very low runoff rates from low but steady rainfall cause very high rates of rill erosion. Susceptibility of the soil to this type of erosion seems much greater on cropland and disturbed sites than on natural, undisturbed lands. Severe ephemeral gully erosion and rill erosion from rainfall on thawing soils occur in many other regions.

Erosion by Piping

Water frequently flows through the soil just below the surface. The soil may contain macropores, other small openings, and channels left by decaying roots, burrowing insects and animals, and other processes. These internal open spaces in the soil may become *pipes*, and pipe flow can erode soil, causing a type of erosion known as *piping*. Usually, the initial diameter of the pipes is quite small, on the order of fractions of an inch (few millimeters) but can be enlarged by erosion to diameters as large as 3 ft (1 m). If a pipe is near the soil surface, the roof of the pipe may collapse, leaving an open rill or permanent, incised gully (Zachar, 1982).

Similar flow occurs where a coarse-textured surface soil lies above a dense soil layer that prevents downward movement of infiltrated water. Water flows downslope through the surface layer of soil, along the top of the dense soil layer, and exits on the lower portion of the hillslope, making this portion of the landscape much more erodible than locations near the top of the hillslope (Huang and Laflen, 1996). A frequent occurrence is increased soil moisture on the lower portions of landscapes and the concentration of soil moisture in hollows. The increased soil moisture and sometimes water seepage makes the soil very erodible at these locations and may accelerate rill and ephemeral gully erosion when runoff from upslope areas passes across these areas (Haung and Laflen, 1996).

WIND EROSION

Differences between Wind and Water Erosion

The flow of wind over the soil surface causes erosion when the erosivity of the wind exceeds the resistance of the soil to erosion. Erosion of soil by wind occurs fundamentally the same way as erosion by water. However, the properties of the respective eroding fluids are very different. Air is much less dense than water. The effect of this difference in density is that wind velocity must be much greater than runoff velocity for wind to have a comparable erosivity to that of water flow over the soil surface. However, the forces holding dry soil in place during wind-erosion events are much less than the forces holding wet soil in place during water-erosion events.

Another important difference is the fluid mechanics of the wind and water flow. A boundary layer develops when a fluid flows over a surface (Figure 3.10). The fluid velocity is zero at the surface and increases with height in the fluid. This boundary layer is much better defined and more stable with wind erosion than with interrill and rill erosion, especially for interrill flow disturbed by raindrop impact. Water flow on overland-flow and concentrated-flow areas has a free surface, which the wind does not have. The velocity of open-channel flow water increases from zero at the

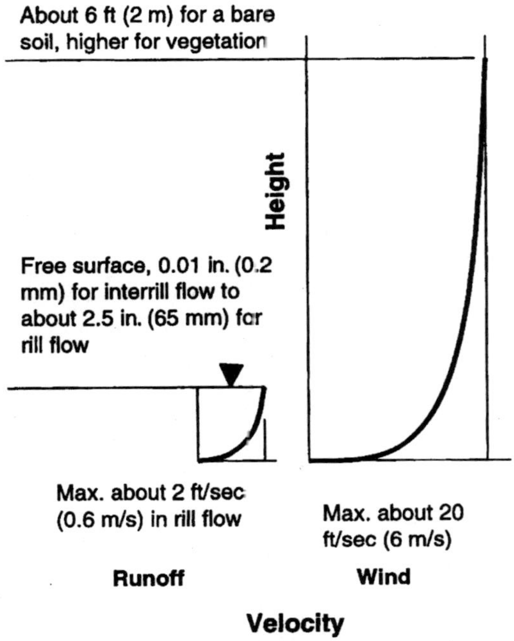

Figure 3.10 Velocity profile for runoff and wind.

soil surface to a maximum velocity flow at or just below the water surface. In wind, the velocity is zero at the soil surface and asymptotically approaches maximum velocity at a height of about 6 ft (2 m) for a bare soil. The ratio of the height of maximum velocity to the diameter of the sediment is much greater with wind erosion than with water erosion. The difference in this ratio affects the difference in the relative importance of sediment transported by suspension in wind compared to the sediment transported by suspension in overland flow and concentrated flow on field-sized areas. Another difference is that the runoff from a rainstorm always flows downslope. However, wind typically blows from a predominant direction during a windstorm, but may change directions between events, although during each season some directions are more probable than others.

Modes of Sediment Transport

Several types of water erosion are defined, but usually only one type of wind erosion is addressed. Wind erosion, however, may be classified on the basis of the sediment transport modes: (1) creep plus saltation, and

Figure 3.11 Deposition of windblown sediment in a fencerow and road ditch. (Courtesy of USDA, NRCS.)

(2) suspension (Chapter 2; Lyles et al., 1983). Windblown sediment transported by creep plus saltation may be deposited very near the source area, along a fence row, in a road ditch, or on a nearby field (Figure 3.11). Sediment transported by suspension, especially fine particles, can be transported great distances as dust before being deposited. *Creep* is sediment transport along the soil surface by skidding, bouncing, and rolling. *Saltation* is sediment transport by the wind lifting particles from the soil surface and moving them downwind in an arc, similar to the trajectory of a projectile (Figure 2.8). These particles reach a height of about 3 ft (1 m) and travel up to 30 ft (10 m) with each saltation. More than 70% of the sediment eroded by wind is transported by creep plus saltation (Lyles et al., 1983).

Modes of Wind Erosion

Wind erosion occurs by two primary mechanisms. Wind exerts forces on the soil sufficient to overcome the resistance of the soil and removes sediment from the soil mass. This type of erosion, *shear erosion*, is similar to scour by water flowing over a soil surface. Once in motion, sediment particles are projectiles that are erosive agents themselves when they strike the soil surface, breaking loose additional soil particles. This type of wind erosion is known as *abrasion*. Not only does windblown sediment abrade the soil and increase erosion, this sediment can damage and kill young plants, causing substantial economic loss for farmers and landscapers.

LINKS BETWEEN WIND AND WATER EROSION

The link between wind and water erosion is minimal. The greatest erosion for the two processes tends to occur in different areas on the landscape, and thus wind and water erosion rates should not necessarily be added to obtain a total erosion rate for an area. Erosion over the landscape is perhaps more uniform when both wind and water erosion occur than when only one process predominates.

One link between wind and water erosion occurs in road ditches and drainage ways where windblown sediment is easily deposited and where runoff flow in these channels easily transports this sediment downstream (Figure 3.11). Had the wind erosion and deposition not occurred, the sediment loads in the channels would have been somewhat less. Another possible link between wind and water erosion occurs off-site, where windblown sediment is deposited in lakes and reservoirs to accumulate with the sediment reaching the impoundments by streams. The total mass of sediment reaching most reservoirs by wind transport is low compared to that delivered by runoff, but the windblown sediment especially can be important if it is eroded from an area contaminated with toxic compounds that are transported with the sediment.

MECHANICAL MOVEMENT OF SOIL

Soil can be moved downslope by the mechanical processes of tillage and mass movement (soil slips). Almost all tillage operations associated with farming lift the surface soil layer. When soil is lifted on a hillslope, gravity causes a net downslope movement of soil. The amount of soil moved by tillage can significantly exceed that moved by interrill and rill erosion (Lindstrom et al., 1992; St. Gerontidis et al., 2001). The amount of soil moved by tillage increases with hillslope steepness and the extent to which tillage lifts the soil. Although not an erosion process itself, soil movement by tillage also reduces soil depth, much like wind and water erosion. Movement of soil by tillage complicates assessment of the impact of erosion in farm fields.

Mass movement in the form of soil slips often occurs on steep hillslopes that receive large rainfall amounts that saturate the upper soil layer and reduce soil strength. When vegetative root networks are insufficient to hold soil in place, a soil mass can slide downslope and into stream channels, where flow transports the sediment downstream. High sediment yields occur in the stream because the mass movement delivered sediment directly to the stream channel (Dunne, 1998). An important sediment source in undisturbed forests is slip that move soil into stream channels.

SUMMARY

Soil erosion is classified on the basis of erosive agent, wind or water. The types of erosion caused by wind are based on the modes of sediment transport: (1) creep plus saltation, and (2) suspension. Medium-sized and coarse particles are transported predominantly by creep and saltation and may be deposited near their source. Fine particles are transported predominantly by suspension, and final deposition of these particles may be thousands of miles (kilometers) from their source.

The classification of erosion caused by water is based on the spatial context in which the erosion takes place. The major types of water erosion are interrill; rill; ephemeral gully; permanent, incised gully; and stream-channel erosion. Rill and interrill erosion occur on hillslopes where runoff occurs as overland flow. Ephemeral gully erosion occurs in the swales of the landscape where topography collects runoff into a few major flow concentrations. The depth of ephemeral gully erosion generally is shallow, so the channels are crossed and easily filled with common agricultural and construction equipment. Permanent, incised gullies occur in the same topographic locations as ephemeral gullies, but these channels are so deep that they cannot easily be crossed or filled. Stream channels, which evolve with and are an integral part of the landscape, transport the runoff and sediment produced on the upland areas of the watershed. The main types of erosion in channels, including rills, are the upstream advance of headcuts and the lateral retreat of sidewalls (bank erosion), each of which can produce large amounts of sediment.

SUGGESTED READINGS

Foster, G. R. 1985. *Understanding Ephemeral Gully Erosion (Concentrated Flow Erosion)*. In: Soil Conservation, Assessing the National Resource: Inventory, National Academy Press. Washington, D.C.

Lyles, L., L. J. Hagen, and E. L. Skidmore. 1983. *Soil Conservation: Principle of Erosion by Wind*. Agronomy Monograph 23. Amer. Soc. of Agron.

Renard, K. G. and G. R Foster. 1983. *Soil Conservation: Principle of Erosion by Water*. In: Dryland Agriculture. Agronomy Monograph 23. Am. Soc. Agron.

Zachar, D. 1982. *Soil Erosion*. Elsevier Scientific Publishing, New York.

4

Erosion Processes

Sediment is eroded on and transported from the landscape by the processes of detachment (separation of soil particles from the soil mass), entrainment (transfer of detached particles from the soil surface to the sediment load), transport (translocation of sediment), and deposition (transfer of sediment from the sediment load to the soil surface) of soil particles by water and wind.

BASIC PRINCIPLES COMMON TO WATER AND WIND EROSION

Water and wind erosion are both unsteady (vary with time) and nonuniform (vary over space). Erosion is unsteady because rainfall intensity and wind velocity that cause erosion vary during almost all rain and wind storms. Also, storms vary from event to event, and the surface conditions that affect erosion change between storms so that the same storm occurring at different times produces different amounts of erosion. Spatial variations in topography, soils, and land use cause corresponding spatial variations in erosion. Also, erosion processes are nonuniform because of physics, even when the factors affecting erosion are uniform. For example, runoff accumulates and erosion rates increase downslope even when rainfall intensity and infiltration rate are uniform. Many processes are similar between water and wind erosion. However, water and wind erosion occur in different environments, dry for wind erosion and wet for water erosion.

The most fundamental principle for understanding soil erosion is that the sediment load is controlled by either the amount of sediment produced by detachment or by the transport capacity of the erosive agents. Water and wind flowing over an erodible soil surface detach soil particles that accumulate as a sediment load along the flow path. A *detachment-limiting condition* occurs where detachment is low on a soil resistant to erosion and sediment load is correspondingly low because detachment does not provide much sediment for transport even though the transport capacity of the flow is greater than the sediment load.

Alternatively, transport capacity is limiting where the supply (availability) of sediment exceeds the flow's transport capacity. One *transport-limiting case* is flow over noncohesive particles. The particles are already detached, and the sediment load is limited only by the ability of the flow to entrain particles and transport them. Two other cases of transport-limiting erosion occur for (1) a system of ridges and furrows on a low grade where deposition occurs in the furrows, and (2) at the bottom of a concave

The analogy of a dump truck transporting soil from a construction site illustrates the concept of detachment or transport capacity limiting the sediment load. The first example shows the sediment load being limited by transport capacity. The dump truck has a capacity of 10 yd^3 (7.6 m^3). A front-end loader with a bucket capacity of 2 yd^3 (1.5 m^3) places soil in the truck, requiring at least five buckets to fill the truck. An absentminded operator places seven buckets of soil on the dump truck and the truck driver drives away. The amount of soil transported is the capacity of the truck, not the amount of soil placed on it. The difference between the amount placed on the truck and the amount transported away becomes spillage over the sides of the truck. Initially, the amount of soil carried away may be greater than the capacity of the truck because the soil was heaped on the truck, but this extra soil falls off the truck along the way. The amount of soil actually delivered to its destination is only slightly greater than the capacity of the truck.

The sediment load also may be limited by detachment. On the next trip, a different front-end loader operator works slowly. The truck driver is impatient and drives off after only three buckets of soil were placed in the truck, which is about 60% of the truck's capacity. On this trip, the amount of soil transported in the truck is determined by the amount of soil actually placed in the truck, not by the capacity of the truck.

slope where deposition occurs (Figure 3.6.). In these cases, the sediment load delivered to a particular location was greater than the flow's transport capacity at that location. Similarly, when the transport capacity of wind abruptly decreases to less than the sediment load at the edge of a field, sediment is deposited in the fencerow or drainage ditch (Figure 3.11).

A second fundamental principle is that the flow's total erosivity is divided between detachment and transport. If the sediment load almost entirely fills the flow's transport capacity, detachment is reduced because most of the flow's erosivity is used to transport the sediment load. Detachment almost ceases as the sediment load approaches the transport capacity (Figure 4.1). The detachment rate equals the detachment capacity when the sediment load is empty, which is illustrated in Figure 4.1 at the point where the uniform flow enters the erodible bed. The detachment capacity is the maximum rate that sediment can be detached for a given set of flow and soil conditions. It is a function of both flow erosivity and soil susceptibility to detachment. As the sediment load fills, the detach-

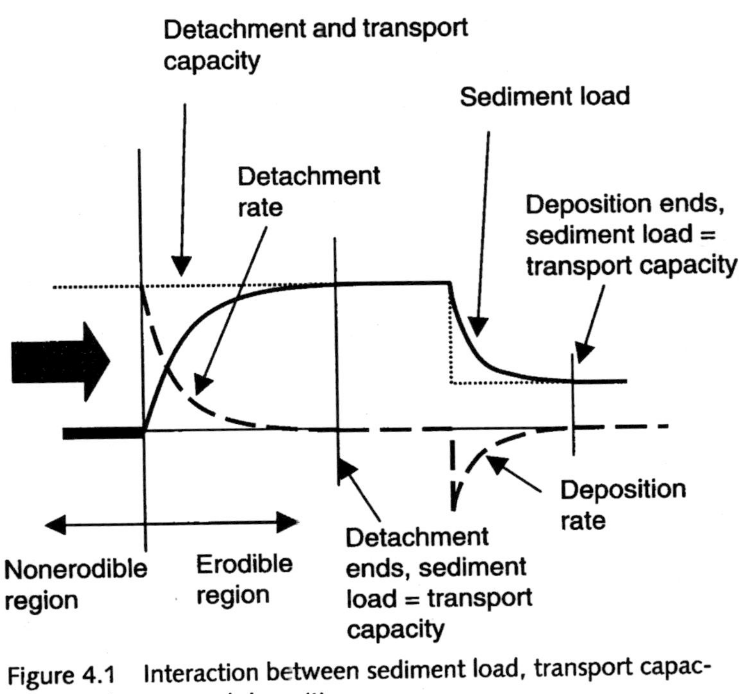

Figure 4.1 Interaction between sediment load, transport capacity, detachment, and deposition.

ment rate decreases along the bed as the sediment load accumulates and fills the transport capacity.

The detachment rate is a fraction of the detachment capacity, depending on the fraction of the transport capacity that is filled by the sediment load. A much longer distance is required for the sediment load to increase to the transport capacity when the soil is resistant to erosion than when the soil is easily eroded. The flow's transport capacity is a function of the flow's erosivity and the transportability of the sediment as determined by the size and density of the sediment particles. If the transport capacity abruptly decreases, as illustrated in Figure 4.1, or the sediment load becomes greater than the transport capacity at a downstream (downwind) location so that the transport capacity is less than the sediment load, the transport capacity becomes overfilled and deposition occurs. The sediment load decreases with distance downstream as deposition occurs, eventually reaching a condition where the sediment load nearly equals the transport capacity for the uniform flow.

The deposition rate is proportional to the difference between the transport capacity and the sediment load, flow velocity, height within the flow that the sediment is transported, and the fall velocity of the sediment particles. *Fall velocity* is the speed with which a sediment particle of a given size and density falls through a column of water or air. When deposition occurs, the coarse particles are deposited rapidly and the sediment load of these particles decreases almost immediately to the flow's transport capacity for these particles (Figure 4.2). A much longer distance is required for the sediment load of the fine particles to decrease to the transport capacity of the fine particles. This selective deposition enriches the sediment load in fine particles and enriches the soil of the depositional surface in coarse particles.

This idealized model of erosion describes wind erosion by creep and saltation, which erode and transport primarily medium-sized and coarse sediment. The erosivity of wind is uniform along the flow path, and the sediment load of the wind is assumed to be empty at the upwind edge of an eroding area. The erosion rate by shear is greatest at the upwind edge of the area and decreases with distance along the flow path as the sediment load increases to the transport capacity for creep and saltation (Stout, 1990; Stout and Zobeck, 1996). A fencerow or road ditch across an eroding area can abruptly cause a decrease in the transport capacity of the wind to a level less than the sediment load. Deposition of the coarse particles occurs at the fencerow and a road ditch as shown in Figure 3.11,

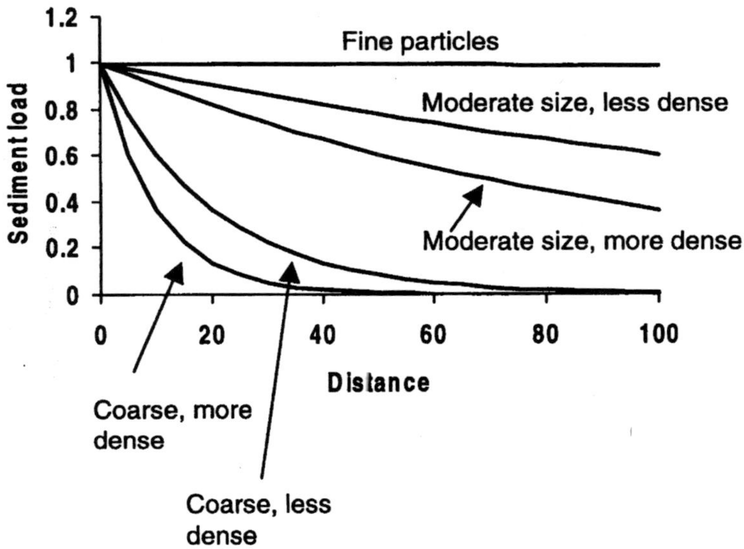

Figure 4.2 Deposition of various particle classes illustrating how deposition is a selective process.

but the fine particles travel much farther downwind before being deposited. The shape of the detachment curve is modified by abrasion, which is discussed later in the chapter.

WATER EROSION

The following discussion of water erosion processes is based on the spatial context of (1) interrill; (2) rill; (3) ephemeral gully; (4) permanent, incised gully; and (4) stream-channel areas (Chapter 3).

Interrill Erosion

Interrill areas are delineated such that all detachment on them is by raindrop impact. Raindrop impact and interrill flow move sediment from the interrill areas to the rill areas. Raindrops striking the soil surface splash soil particles in all directions, but the net movement is downslope and increases as interrill steepness increases. Most of the sediment delivered from interrill areas to rill areas, however, is transported by the thin flow on the interrill areas rather than by rainsplash (Meyer et al., 1975b; Mutchler and Young, 1975). Sediment delivery from interrill areas to rills

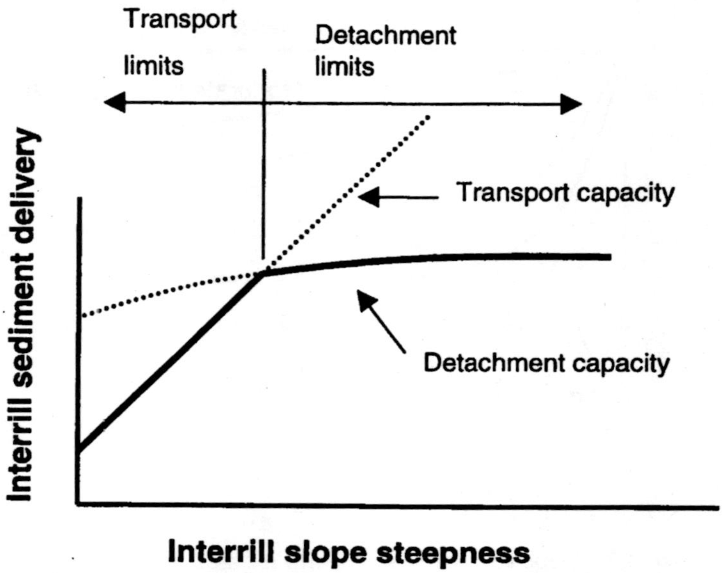

Figure 4.3 Detachment and transport capacity limits on interrill sediment delivery.

is governed by the principle that the sediment load is limited by particle detachment or by transport capacity on the interrill areas, as described earlier by the analogy of the dump truck. Detachment limits interrill sediment delivery in most cases, but interrill transport capacity can limit sediment delivery when interrill steepness is low. Figure 4.3 illustrates the principle of interrrill sediment delivery being limited by either detachment by raindrop impact or by transport capacity of the interrill runoff.

Detachment by Raindrop Impact

A raindrop striking a soil surface exerts intense, though brief fluid forces on the soil (Nearing et al., 1987), and creates an impact crater when striking a layer of noncohesive soil (e.g., loose sand). Initially, flow in the crater entrains sediment so that the sediment load accumulates in the radiating outflow. As flow continues outward, the transport capacity within a single raindrop impact decreases to less than the sediment load, so that deposition begins and continues until the flow ends. The size of the crater created by the raindrop depends on the size and impact velocity of the raindrop, size and density of the soil particles, and the depth of water on the soil surface (Mutchler, 1970, Mutchler and Young, 1975). In general, the size of the crater, indicative of the amount of sediment that is moved, increases with raindrop diameter and impact velocity. A raindrop that is small or that strikes the soil at a low velocity will move little or no sediment. Even large raindrops cannot move large, dense particles such as gravel.

A thin depth of water on the surface increases the amount of sediment moved by raindrops. However, as this depth increases, a water depth is reached where the raindrop forces applied to the soil begin to decrease. A water depth greater than three times the drop diameter essentially eliminates forces applied to the soil by raindrops (Mutchler and Young, 1975).

A cohesive soil reacts differently to raindrop impact than does a noncohesive soil (Figure 4.4). Craters are not developed on a cohesive soil because of internal bonding among soil particles. All of the raindrop forces on a noncohesive soil are expended in entraining and transporting sediment, but on a cohesive soil most of these forces are expended in detaching soil particles. The amount of sediment moved on a noncohesive soil by a single raindrop impact is transport-limited but is detachment-limited on a cohesive soil. The amount of sediment moved by a single raindrop striking a cohesive soil is much less than the amount of sediment moved on a noncohesive soil.

Erosion generally is considered to be selective, where fine particles are removed and coarse particles are left behind to enrich the soil in sand and gravel gradually, through time. A close examination of erosion processes

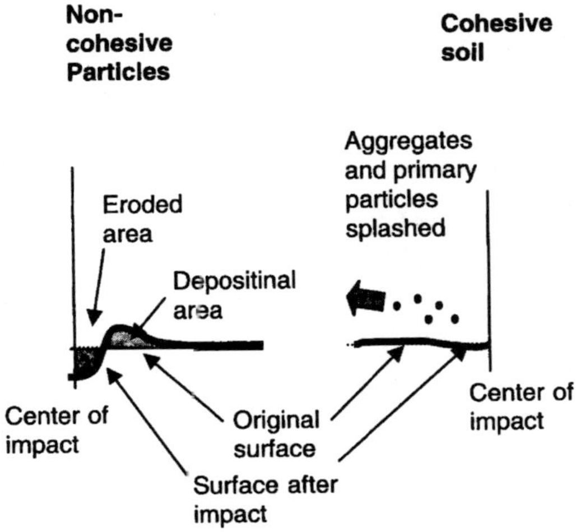

Figure 4.4 Raindrop impact on noncohesive and cohesive soils.

reveals how this selectivity occurs. Detachment by raindrop impact on a cohesive soil is nonselective. The forces of raindrop must overcome the soil's internal bonding forces to cause particle detachment. When the bonding fails during raindrop impact, particles of all sizes and densities comprising that soil are produced. A raindrop cannot "reach" into a cohesive soil and selectively remove certain particles without breaking bonds that release all particle classes. The force required to move large particles such as gravel is greater than the force required to detach particles from a cohesive soil. The detached small particles are moved, but the gravel is left behind to armor the surface. Entrainment rather than detachment becomes the limiting factor when large, dense particles are present on and in the soil.

Erosivity

Total detachment on an interrill area is the sum of the detachment caused by each raindrop or waterdrop falling from a canopy. The forces applied to the soil by waterdrops, whether raindrops or drops from the canopy, increase with impact velocity and the mass of the drops. Because the forces applied to soil by a raindrop are difficult to measure (Nearing et al., 1987), the usual measure for erosivity of a single raindrop is its impact energy, which combines the mass of the raindrop and the square of its impact velocity (Sharma et al., 1993). A measure of total erosivity of a storm, therefore, is the sum of the impact energies of all raindrops during the storm. Unit energy of a rainfall energy per unit of rainfall is a function of rainfall intensity (Figure 4.5). This relationship is used to compute total storm energy by taking into account the amount of rainfall that occurs at different intensities during the storm.

The theory and measurement of detachment by individual raindrops indicate that storm energy alone should be sufficient for estimating the total amount of sediment detached by raindrop impact during a rainstorm, but such is not the case. The raindrops and the thin interrill flow interact so that the erosivity of the storm is greater than the sum of the erosivity of the individual drops. The product of total storm energy and peak rainfall intensity is a much superior measure of single-rainstorm erosivity than is energy alone (Chapter 2; Wischmeier, 1959). The average rainfall intensity during the 30 minutes with the maximum rainfall amount is widely used to represent the effect of rainfall intensity on erosion, based on experimental research, but the best measure of peak rain-

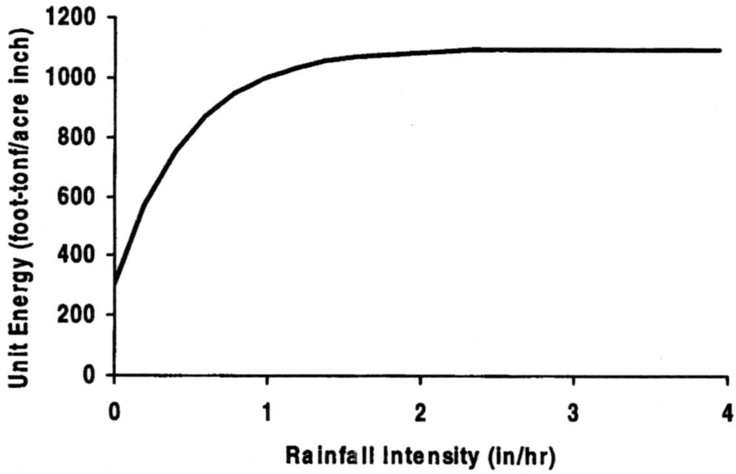

Figure 4.5 Unit rainfall energy as a function of rainfall intensity. Multiply intensity by 25.4 to obtain mm/h and multiply unit energy by 2.638×10^{-4} to obtain MJ/(ha·mm).

fall intensity actually varies with both storm and soil characteristics (Chapter 5).

Total detachment by raindrop impact during a storm can also be determined from detachment rate and how it varies with rainfall intensity during the storm. At any particular time interval during a storm, the detachment over that time interval is the average detachment rate for the interval times the duration of the interval. Total detachment for the storm is the sum of the detachment for each interval. The detachment rate by raindrop impact for each interval varies approximately with the square of rainfall intensity for that interval (Foster, 1982; Meyer, 1981).

Canopy cover, ground cover, and ponded water in depressions reduce the erosivity of raindrops. The effect of canopy on erosivity of raindrops is described in Chapter 2. The effect of ground cover, for example, on interrill erosion is shown in Figure 4.6 (Khan et al., 1988; Lattanzi et al., 1974; McGregor et al., 1990). Water ponded in depressions associated with random roughness eliminates the erosive forces of raindrop when the water depth is greater than three times the drop diameter.

Soil Resistance

Detachment on interrill areas is a function of the soil's resistance to detachment by raindrop impact. Theoretical studies and experimental re-

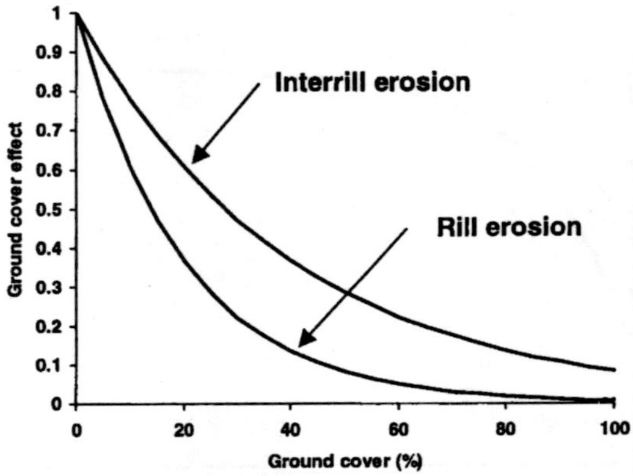

Figure 4.6 Effect of ground cover on interrill and rill erosion.

search show that detachment is related to internal soil strength measured on the scale of the raindrops and soil aggregates produced by detachment (Al-Durrah and Bradford, 1982). Measuring the internal bonding forces in soil and the forces applied by raindrop impact is impractical because these forces are difficult to measure and they vary over the soil surface. Instead, experimental methods are used to determine empirical measures of interrill soil erodibility. Measured interrill sediment delivery from natural rainfall or artificially applied rainfall, where detachment is limiting rather than transport capacity, are plotted against rainfall erosivity, much like Figure 2.3 (Free, 1960, Meyer, 1981). The slope of the line in Figure 2.3 is an empirical measure of soil erodibility. This erodibility is not a soil property like soil texture but is defined by the variables used to compute rainfall erosivity, the soil conditions under which the measurements are made, and the measurement technique (Bradford and Huang, 1993). Actual field conditions modify soil erodibility from the value determined from the base or standard research conditions used to measure erodibility. The effect of those factors on erodibility are described in detail in Chapter 2.

Interrill Sediment Transport Capacity

Rainsplash and interrill flow are the two modes of sediment transport on and from interrill areas. Rainsplash transport is the movement of sediment particles through the air when a raindrop strikes the soil surface and launches sediment during detachment or launches previously detached sediment laying loose on the soil surface. Rainsplashed sediment can reach heights up to about 3 ft (1 m) and horizontal distances of about 3 ft (1 m), but the size and amount of sediment splashed per unit height decreases with height above the soil (Foster et al., 1985a). The amount of rainsplashed sediment decreases as the size and density of the sediment increases, but increases with rainfall intensity. The amount of rainsplashed sediment is greater when previously detached sediment is loose on the soil surface than when the impacting raindrops first must detach particles from the soil as part of a single impact.

Almost all of the sediment delivered from interrill areas is by interrill flow (Meyer et al., 1975b, Mutchler and Young, 1975). The sediment-transport capacity of interrill flow results from the combined action of the interrill flow and raindrop impact on the flow. Without the raindrop impact, interrill flow would transport only very fine sediment. The contribution of raindrop impact to the transport capacity of the interrill flow increases as rainfall intensity increases. However, the contribution of raindrop impact diminishes as the flow rate and interrill slope steepness increase (Moss et al., 1979). Altogether, the interrill transport capacity increases as the interrill flow rate and the interrill steepness increase.

The size and density of the sediment particles is the other major factor that determines interrill sediment-transport capacity. Interrill transport capacity is less for sand-sized particles than for silt- and clay-sized particles (Foster, 1982). Large dense particles such as gravel are not transported from interrill areas because the interrill flow and impacting raindrops cannot move these particles.

At low interrill slopes, the amount of sediment produced by detachment exceeds the transport capacity of interrill flow, which causes deposition. Coarse particles are left on the surface while the fine particles are transported from the interrill area (Figure 4.8). This selective process enriches the sediment load in fine particles and enriches the soil surface in coarse particles relative to the original soil (Alberts et al., 1980; Foster, 1982; Hairsine and Rose, 1991; Monke et al., 1977; Proffitt et al., 1991).

When transport-limiting conditions exist under steady rainfall, a layer of loose sediment particles accumulates on the interrill surface and becomes sufficiently deep to dissipate the erosive forces of the raindrops (Hairsine and Rose, 1991; Profitt et al., 1991). Detachment eventually decreases to a rate that just balances the transport capacity (Figure 4.7). Hereafter, deposition and particle selectivity ends, so that the composition of the sediment leaving the interrill areas is the same as the soil being

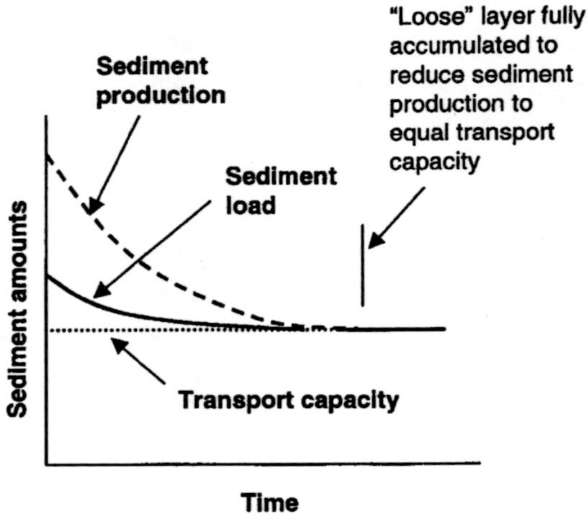

Figure 4.7 Effect of accumulation of a loose sediment layer on interrill erosion.

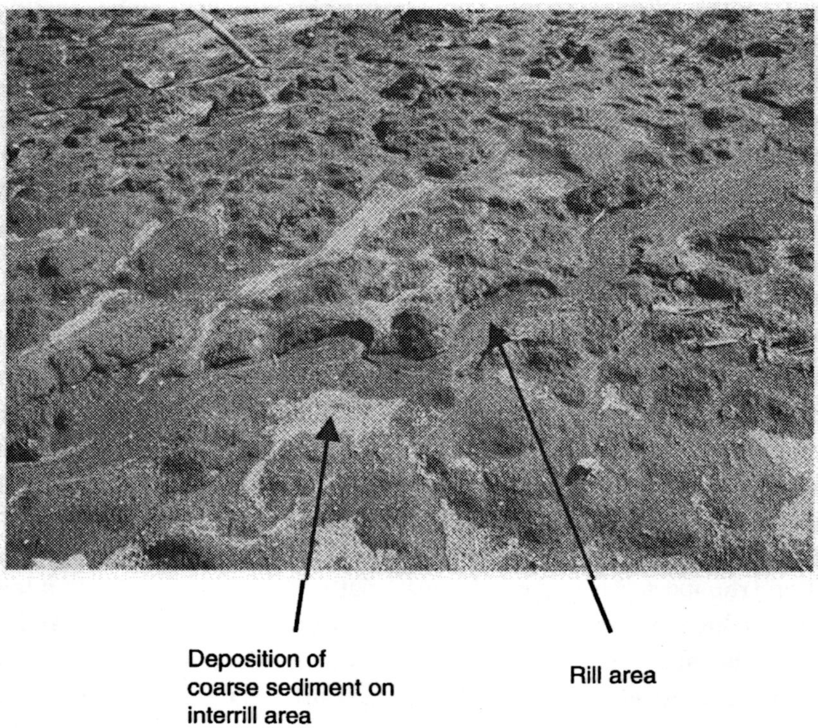

Figure 4.8 Transport capacity limits interrill sediment delivery resulting in coarse sediment being deposited.

eroded, assuming that raindrop erosivity is sufficient to move all particle classes in the soil.

This effect is most obvious on very coarse-textured and gravelly soils that have not recently been disturbed. The sand and gravel particles are too large and dense to be transported by interrill flow and raindrop impact, and through time, the fine particles are removed. A loose layer of very coarse particles develops on the soil surface to reduce detachment in proportion to the fraction of the soil surface covered by these coarse size particles. This selective transport by particle size and density creates an armor of nontransportable particles on the soil surface (Hairsine and Rose, 1991; Profitt et al., 1991).

The same process develops an armor layer on gravel roads. Routine grading for maintenance leaves a layer of mixed-textured, loose soil particles on the surface. The easily transported particles are removed during the first storm event, but the gravel and other particles that are too large and dense to be transported are left behind to armor the road (Elliot, 1999). Erosion decreases significantly as the armor layer develops. The same process can develop an armor layer on agricultural soils, but the rate of armoring is slower on these soils than on gravel roads because of differences in the amount of gravel in a unit volume of soil and mixing the gravel in soil by mechanical disturbances.

Interrill sediment-transport capacity is also related to flow velocity (Foster, 1982). Hydraulically rough soil surfaces reduce flow velocity and hence interrill transport capacity. Thus, transport capacity is much less on a rough interrill surface or one covered with plant litter than on a bare, smooth surface. Deposition occurs on rough interrill surfaces (Figure 4.7). During rainfall and runoff events, the soil that protrudes above the ponded water in the depressions is exposed to raindrop impact that detaches substantial amounts of sediment. The transport capacity of the interrill flow through the depressions is very low because of the slow flow velocity. Detachment produces more sediment than the interrill flow can transport, causing deposition in the depressions. The sediment deposited in the depressions is coarser than the sediment leaving the interrill area, which enriches the soil in coarse particles and demonstrates again how erosion can be a selective process. As the depressions fill with deposited sediment and the surface roughness decreases, the transport capacity increases, which increases interrill sediment delivery, reducing deposition and the selective removal of the fine sediment. Through time, interrill sediment delivery matches interrill detachment and the composition of the sediment delivered from the interrill areas matches the composition of the soil being eroded (Foster, 1982; Meyer et al., 1980). A high degree of roughness can cause about 70 percent of the sediment detached by raindrop impact to be deposited in depressions (Foster, 1982). The fraction deposited and the degree that deposition enriches the interrill sediment delivery in fines depends on the characteristics of the detached sediment. The degree of enrichment of fines increases as the fraction deposited increases.

Just as canopy cover reduces the erosivity of raindrops for detachment, it also reduces interrill sediment-transport capacity because raindrop impact enhances interrill transport capacity. Ground cover also reduces interrill transport capacity in much the same way that random roughness decreases interrill transport capacity. A difference is that ground cover does not provide depressions to store sediment as does random roughness.

Rill Erosion

The rill areas receive the runoff and sediment delivered from interrill areas. Rill erosion is caused by flow in rill areas (Chapter 3). Interrill and rill areas combine to make up the overland-flow portion of the hillslope.

Basic Principles

The basic principle of detachment or transport limiting also applies to rill areas. The sediment load leaving a uniform, heavily rilled hillslope is limited by soil detachment. If the sediment delivery from the interrill areas adjacent to the rill areas exceeds the transport capacity of the flow in the rill areas, deposition occurs on the rill areas. Also, a decrease in steepness or a dense vegetative strip on a hillslope can reduce the transport capacity to cause deposition.

Several other basic principles apply to the detachment and deposition processes that occur in rill areas. First, when erosion by flow (detachment) occurs, it erodes an incised channel. Second, when deposition by flow occurs, the flow is broad and shallow rather than concentrated during rill erosion (Figure 3.6). Third, stable erosional features of headcuts and channel cross sections can only be maintained by upstream movement of the headcuts and downward movement of the cross sections (Foster, 1982). These principles and characteristics of rill erosion apply on both noncohesive and cohesive soils. The internal soil strength of cohesive soils allows incised channels with nearly vertical sidewalls to develop, which cannot develop in noncohesive soils because of the lack of soil strength. Another difference between noncohesive and cohesive soils is that the particles in noncohesive soils are already detached, and thus detachment on noncohesive soils really is entrainment.

Rill erosion creates rills (Figures 3.5 and 3.6). Rill erosion occurs by the upslope advance of headcuts, incision of rills moving downward into the soil, and the retreat of rill sidewalls. Headcuts are local areas of very intense erosion and can vary in shape from overfalls to chutes (Figure 4.9). The spacing and geometry of the headcuts and the rate that they move upslope are a function of soil strength, soil erodibility, slope steepness, and flow rate (Foster, 1982; Meyer et al., 1975a).

Another way that flow erodes incised channels results from the nonuniform distribution of the flow's shear stress around a rill boundary, which causes nonuniform erosion around the rill (Foster et al., 1984). The shear stress and erosion rate are greatest in the middle of the rill. Rill erosion develops a channel with a specific geometry that moves downward into the soil (Figure 4.10). The geometry of the channel and the erosion rate are functions of soil strength, soil erodibility, slope steepness, and flow rate (Foster, 1982).

Rills erode downward into cohesive, uniform soil without widening. The apparent widening caused by erosion results from the mechanical slough-

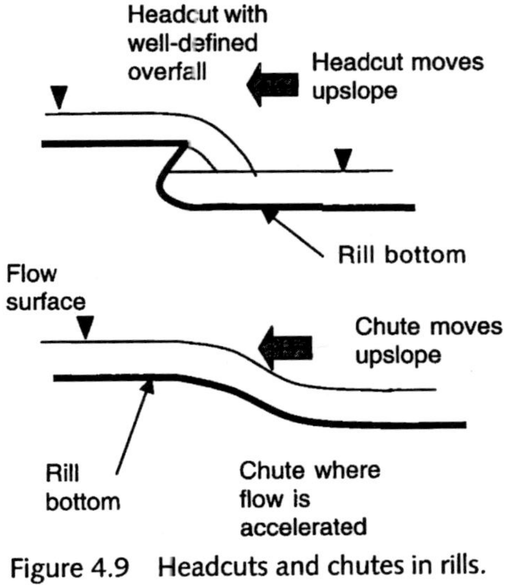

Figure 4.9 Headcuts and chutes in rills.

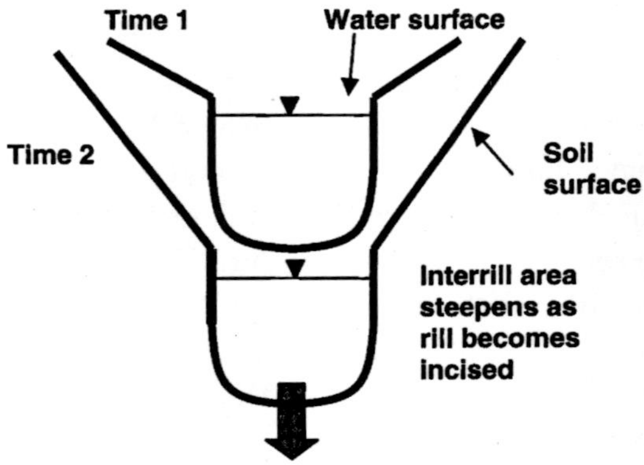

Figure 4.10 Equilibrium rill eroding downward in uniform soil.

ing of the sidewalls and accelerated interrill erosion, illustrated by increased interrill sideslopes in Figure 4.10. Because soil strength usually cannot maintain a vertical rill sidewall, soil collapses into the rill and is "cleaned out" by the flow. The slumping of the rill sidewall steepens the adjacent interrill areas and increases interrill erosion, which reshapes the interrill areas. These effects can be seen most clearly on steep construction hillslopes, where rill erosion is very intense (Figure 4.11).

Most soils are naturally layered (Appendix A). Tillage, however, creates a surface layer that can easily be eroded, while the untilled soil just beneath the tilled layer is more resistant to erosion and acts as a nonerodible layer (Foster, 1982). Flow erodes through the erodible tilled layer to the untilled layer, which restricts downward incision of the rill. When a rill reaches a nonerodible layer, flow begins to undercut the rill sidewalls and the rill widens. The widening continues until the shear stress of the flow at the nonerodible layer matches the critical shear stress of the soil (Foster, 1982).

The width of the rill as it erodes in a uniform soil is called the *equilibrium width*, and the maximum width caused by the nonerodible layer is called the *final width*. Both widths are a function of discharge rate, slope steepness, and critical shear stress of the soil (Foster, 1982). When the rill reaches the nonerodible layer, it begins to widen and the erosion rate decreases to zero as rill width approaches the final width. Once the rill reaches a nonerodible layer, the erosion caused by a particular flow rate depends on how close the existing rill width is to the final width for the new flow rate. No erosion will occur if the existing final width exceeds the final width for the new flow rate. Although these descriptions apply to a

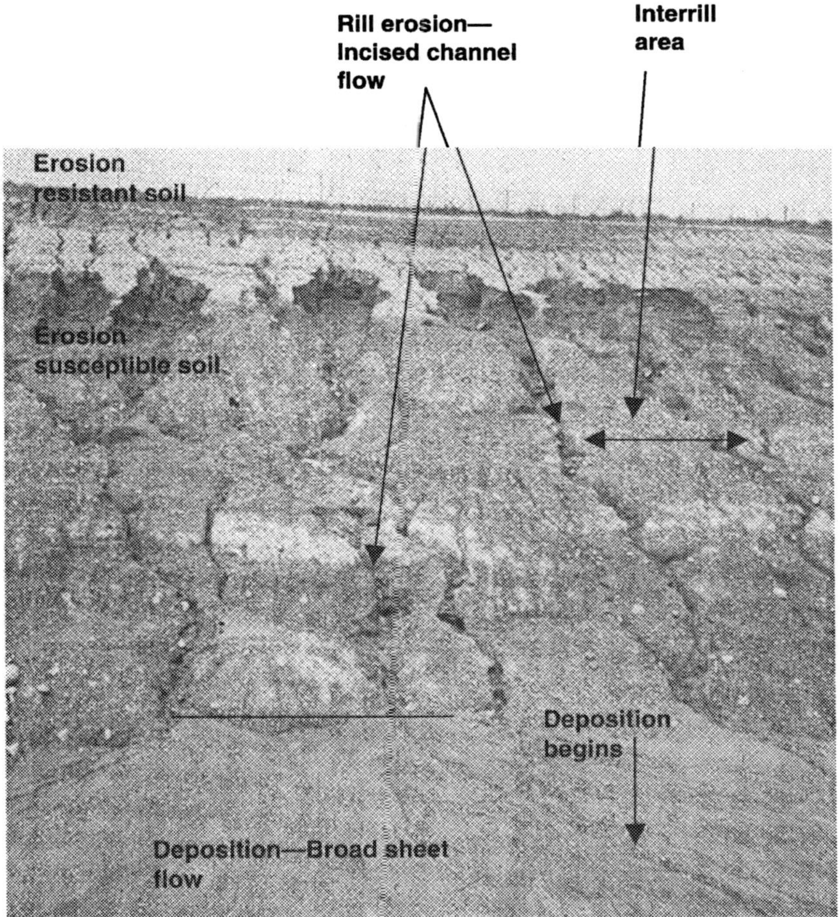

Figure 4.11 Intense rill erosion on a construction site.

steady flow, the principles also apply to unsteady flow, which is typical of runoff during rainstorms. For example, rill erosion will be much higher for a rainstorm that occurs immediately after tillage than if the same rainstorm occurs after rills have been developed above a nonerodible layer.

Erosivity of Rill Flow

A typical relationship for rill erosion as a function of flow hydraulics is illustrated in Figure 4.12, where the rill-erosion rate is a function of the temporal and spatial average shear stress (Foster et al., 1977). The slope of the line is a rill-erodibility factor and the abscissa intercept on the graph

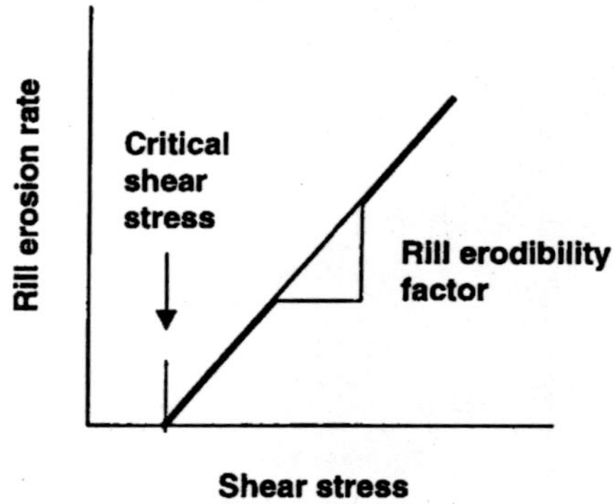

Figure 4.12 Relation of rill erosion to shear stress.

is a critical shear stress. Of course, for erosion to occur, the shear stress of the flow must exceed the critical shear stress of the soil. Experiments to determine values for these parameters should be conducted under field conditions because typical field temporal and spatial variability affect empirical values of critical shear stress. For example, the average shear stress can be less than the true critical shear stress of the soil, but concentrated shear stress at headcuts and chutes or bursts of intense shear stress as a part of turbulent flow can exceed critical stress and cause localized erosion at particular points in time and space (Foster et al., 1984).

The shear stress acting on the soil often is used to represent the erosivity of flow. This variable is proportional to the product of flow rate and slope steepness to the two-thirds power (Foster et al., 1977). Other variables can be used to represent flow erosivity, but all of them basically involve a product of flow rate and slope steepness. When water flows over a soil surface, it exerts a force on the surface over which it passes. When hydraulic roughness is present in the form of random roughness, plant stems, ground cover (including last year's crop residue), plant litter, naturally occurring gravel, or applied erosion-control materials, the total shear stress exerted by the flow is distributed among these hydraulic-roughness elements and the soil particles. Only the portion of the total shear stress that is exerted on the soil particles causes detachment and sediment transport (Foster, 1982). Increasing total hydraulic roughness increases the total shear stress applied by the flow. However, the proportion of the total shear stress that is applied to soil particles decreases more rapidly than the total shear stress increases. The result is that hydraulic roughness reduces the shear stress applied to soil particles, which in turn reduces the detachment and sediment-transport capacity of runoff and hence rill erosion. The ratio of shear stress applied to the soil particles by these forms of hydraulic roughness to the shear stress applied to the soil particles for a smooth, bare surface equals the square of the ratio of the flow velocity with the roughness to the flow velocity for a smooth, bare surface. For example, the ratio of the flow velocity with a heavy mulch cover to the flow velocity for a smooth, bare surface might be about 0.3 (Foster and Meyer, 1975). The square of this ratio is 0.09, which means that the shear stress applied to the soil particles is about 0.1 with mulch to that without mulch. The rill-erosion rate is approximately related to the ratio of these shear stresses to the 3/2 power, which means that rill erosion with the mulch is $0.09^{1.5} = 0.03$ (Foster, 1982). That is, this amount of mulch would reduce rill erosion by 97%. These effects of mulch on rill erosion are illustrated in Figure 4.6.

Buried material in the soil, such as roots and large pieces of crop residue, also reduce the shear stress applied to the soil as rill erosion exposes these materials (Brown et al., 1989; Franti et al., 1996; Van Liew, and Saxton, 1983). Thus, these buried materials reduce rill erosion by reducing the stress that the flow applies to soil particles. These materials reduce rill erosion in another way, by acting as miniature grade-control structures, provided that the buried material is sufficiently large and dense to resist displacement itself.

Soil Resistance to Detachment

Rill erosion is a function of the forces applied to the soil by rill flow in relation to the forces within the soil that resist detachment. Soil strength measured on the scale of the aggregates should be a fundamental measure of soil resistance (Al-Durrah and Bradford, 1982). However, the factors that affect rill soil erodibility are very complex and are typically determined by empirical studies (Chapter 2).

Rill Transport Capacity

Almost all of the sediment detached on overland-flow areas is transported downslope by flow in the rill areas. The portion of the flow's total shear stress that acts on the soil particles along with the sediment-particle diameter and density determine the sediment-transport capacity of flow in the rill areas. Transport capacity increases as the flow rate and slope steepness increase. Ground cover, random roughness, and plant stems reduce the transport capacity in the same way that they reduce detachment capacity (Foster, 1982).

The flow must transport a range of particle sizes and densities, from very small particles such as clay-sized particles that are less than 1 μm in diameter, to sand-size particles greater than 1000 μm in diameter. Similarly, particle densities (specific gravities) range from less than 1.4 to 2.65 (Foster et al., 1985b). The large and dense particles are much more difficult to transport than the small and less dense particles. Flow can entrain the easily transported fine and less dense particles in previously detached (loose) sediment on the soil surface, leaving the coarse and dense particles to armor the surface. However, the flow cannot "extract" fine and easily transported particles from a cohesive soil. Detachment on a cohesive soil produces the full suite of particle classes. Also, the critical shear stress required for detachment is much greater than that for transport of sand-size and smaller particles (Foster, 1982). The result is that detachment and deposition do not occur simultaneously, and the selectivity of rill erosion, as for interrill erosion, occurs through entrainment, transport, and deposition processes rather than by detachment processes.

Rill Area Deposition

Deposition is a selective process such that coarse and dense particles are deposited first and fine and less dense particles are transported farther downstream. Deposition occurs when and where the sediment load becomes greater than the transport capacity of the flow. The deposition rate is proportional to the difference between the transport capacity and sediment load, the fall velocity of the sediment, and the inverse of the product of flow depth and velocity (flow rate per unit width) (Foster et al., 1985c). Coarse and dense particles are readily deposited, so the sediment load of these particles closely tracks the flow's transport capacity for these particles. Conversely, deposition of the finest and least dense particles is very slow, and the sediment load of these particles closely tracks the amount of the particles added to the flow by upstream detachment in rill areas plus interrill sediment delivery.

Interaction of Interrill and Rill Areas

Sediment yield measurements, plotted in Figure 4.13, at the downstream end of a simple ridge–furrow system illustrates how interrill and rill areas

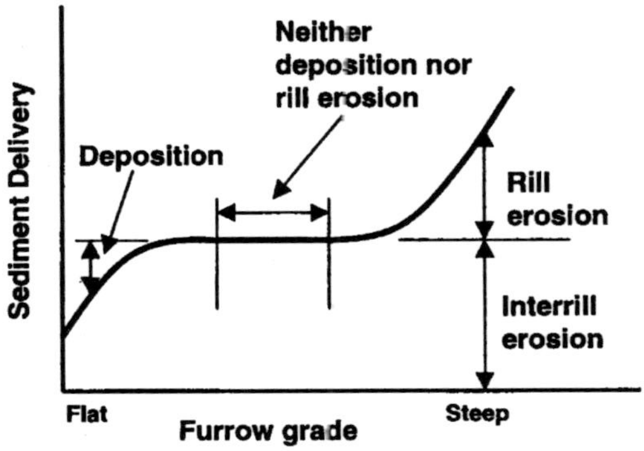

Figure 4.13 Sediment yield from a ridge–furrow system as a function of furrow grade.

interact. When the furrow grade is very flat, deposition occurs in the furrow (rill area), resulting in sediment delivery from the furrow being less than interrill sediment delivery to the furrow and the sediment being finer than the soil texture of the interrill areas. The degree of enrichment in fine sediment delivered from the furrows indicates the degree of deposition in the furrow.

An increase in the furrow grade increases the transport capacity of flow in the furrow, which decreases deposition in the furrow to the point that deposition ends and sediment delivery from the furrow equals sediment delivery from interrill areas. As deposition decreases, the composition of the primary particles delivered from the furrow becomes the same as the sediment delivered from the interrill areas.

Sediment delivery from the furrow equals interrill sediment delivery as furrow grade increases through the midrange of furrow grades, which indicates that no deposition or detachment is occurring in the furrow. The sediment-transport capacity of flow in the furrow exceeds its sediment load, but no rill erosion occurs because the shear stress of the flow is less than the critical shear stress of the soil. At a particular furrow steepness, however, the flow's shear stress exceeds the soil's critical shear stress, and detachment by flow (rill erosion) begins in the furrow. The composition of the primary particles from the furrow after rill erosion begins is the same as that of the soil, which demonstrates that detachment processes on interrill and rill areas are nonselective.

This same fundamental behavior occurs on hillslopes, like the complex shaped hillslope in Figure 2.4*a*, where the upper section is convex and the lower section is concave. Figure 4.14 is detailed presentation of Figure

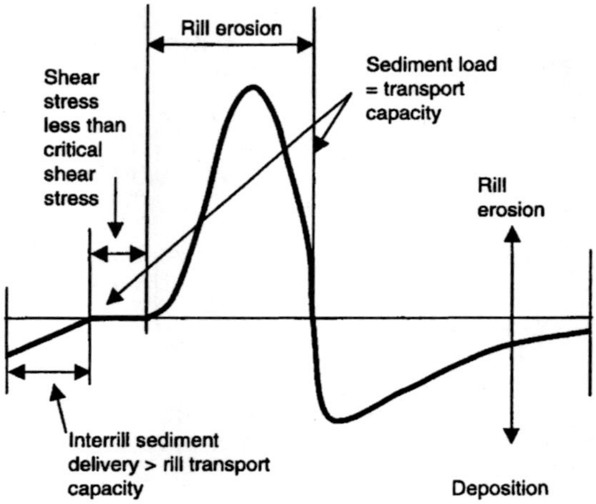

Figure 4.14 Erosion processes in rill areas along a complex hillslope profile.

2.4*b*. The erosion processes illustrated in Figure 4.14 represent those on the hillslope shown in Figure 3.6. On the uppermost portion of the slope profile, where steepness is nearly flat, the transport capacity in the rill areas is low and deposition occurs because interrill sediment delivery exceeds the transport capacity. As steepness increases along the slope profile, deposition decreases as the transport capacity increases because of increases in both flow rate and steepness.

A location is reached where deposition ends because the transport capacity exceeds the sediment load. A segment of the slope may exist after deposition ends where no detachment by flow occurs because the flow's shear stress acting on the soil is less than the critical shear stress of the soil. However, the flow's shear stress increases with distance along the slope as flow accumulates and steepness increases so that the soil's critical shear stress is exceeded and rill erosion begins.

Slope steepness increases to a maximum and begins to decrease. Because the transport capacity is a function of the product of distance along the slope (which represents the effect of flow accumulation on transport capacity) and slope steepness, the transport capacity reaches a maximum beyond the location of maximum steepness. However, at some location, the decrease in steepness overcomes the increase in flow (distance) and the transport capacity begins to decrease toward the end of the slope. Even though rill erosion decreases along with transport capacity, the sediment load continues to increase from interrill sediment delivery and rill erosion.

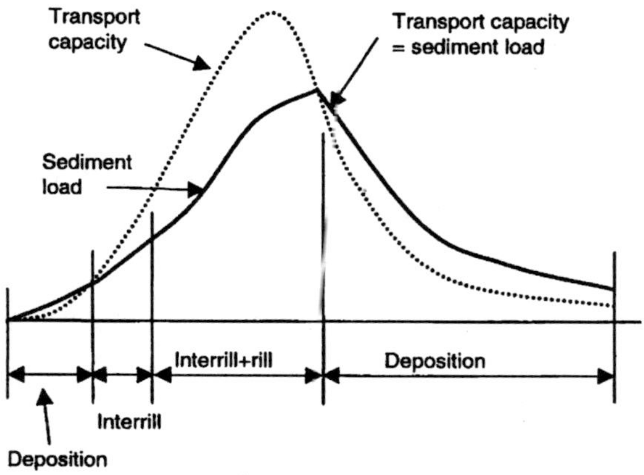

Figure 4.15 Sediment load along a complex hillslope profile.

Figure 4.15 is a plot of sediment load along the slope. The slope of the curve for sediment load is the same as the plot of detachment and deposition in Figure 4.14.

A location is reached where the sediment load equals the transport capacity; thereafter, deposition begins and continues for the remainder of the slope. The flow shifts from concentrated rillflow to a broadsheet flow that spreads the deposition across the slope base (Figures 3.6 and 4.11). The coarsest and densest particles are deposited first on the slope at the upper edge of deposition; the finer and less dense particles are transported farther downslope before being deposited. The composition of the sediment load that leaves the slope is finer than that of the soil producing the sediment, and the sediment load leaving the slope is much less than the total amount of sediment produced by detachment on the slope (Davis et al., 1983; Lu et al., 1988).

Another important example of interactions among detachment, sediment transport, and deposition is the deposition caused by a strip of dense vegetation below a slope segment where erosion rates are high (Figure 4.16). The hydraulic roughness of the dense vegetation greatly reduces flow velocity and thus the transport capacity, so that the transport capacity at the upper edge of the vegetation is much less than the incoming sediment load, resulting in deposition (Dabney et al., 1993). The transport capacity decreases abruptly at the upper edge of the vegetation and increases with distance within the vegetation. If the vegetation strip is sufficiently wide, deposition will end within the strip, and erosion, although

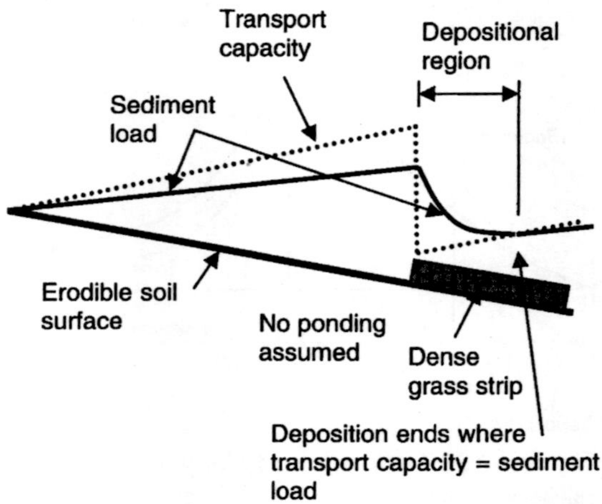

Figure 4.16 Effect of grass strip on transport capacity and sediment load.

small, will occur on the lower portion of the strip. The sediment delivered from the strip is much less than the sediment load that reaches the upper end of the strip, and the sediment leaving the strip is much finer than the sediment that reaches the strip. A strip width wider than the location where deposition ends does not significantly increase sediment trapping within the strip.

This representation of the vegetation strip is simplified from actual field conditions. The hydraulic roughness of the strip causes flow to pond in backwater just upstream of the upper edge of the strip. Flow velocity in the backwater is very low, and thus the transport capacity is much reduced, which induces deposition in the backwater. In fact, more deposition occurs in the backwater than within the strip itself until the backwater areas become filled with sediment (Foster, 1982).

The upper edge of deposition advances upslope rapidly as deposition occurs in the backwater, so that the area of reduced transport capacity is much larger than the initial backwater area. Continued deposition steepens the depositional area and increases the transport capacity, which increases sediment transport into the vegetation. As sediment accumulates in the vegetation, the transport capacity in the upper part of the vegetative strip increases and sediment is transported farther into the vegetation. Eventually, the strip is filled with sediment and the strip no longer traps sediment (Foster, 1982).

Concentrated Flow

Concentrated-flow areas end overland-flow slope lengths (Chapter 3). Most of the concentrated flow areas occur in swales that are parts of natural landscapes, and if a landscape, such as a reclaimed surface mine, is not constructed with sufficient drainage channels, concentrated flow areas develop as a part of the landscape's evolution. Concentrated-flow channels, such as diversions and terraces, are constructed on overland-flow areas to intercept overland flow, preventing excessive rill erosion.

Ephemeral Gullies

Ephemeral gullies occur in concentrated-flow areas located in the swales of the landscape (Chapter 3). The processes that occur in ephemeral gullies are the same as for rill erosion, but the effect of a nonerodible layer is far more important for ephemeral gully erosion than for rill erosion. Nonerodible layers are usually sufficiently deep relative to the erosion rate in rills that the rills are much slower in reaching a nonerodible layer than an ephemeral gully. The depth to the nonerodible is the same for both rills and ephemeral gullies, but the downward erosion rate in the ephemeral gully is an order of magnitude or more greater than the downward erosion rate in rills. The effect of the nonerodible layer in an ephemeral gully means that ephemeral gully erosion rate for a particular storm depends on the amount of previous erosion. If a large storm occurs where tillage has recently filled the ephemeral gully, erosion will be far greater than if the same storm occurs after several large storms. Similarly, a small storm may cause considerable ephemeral gully erosion if it occurs after a recent tillage operation, but no erosion if it occurs after a large storm that causes much erosion. The bottom of the concentrated flow area in Figures 1.1 and 3.7 is determined by a nonerodible layer.

For a steady discharge rate, an ephemeral gully will erode to a final width that depends on discharge rate, grade of the channel, and soil conditions. Figure 4.17 illustrates how erosion rate decreases as channel width increases in an ephemeral gully for a steady flow. Although flow is not steady during a runoff event, a "representative" flow rate can be assumed for each event to illustrate how ephemeral gully erosion varies between storms. The amount of erosion that occurs for an event depends on the difference between the ephemeral gully width before the storm to

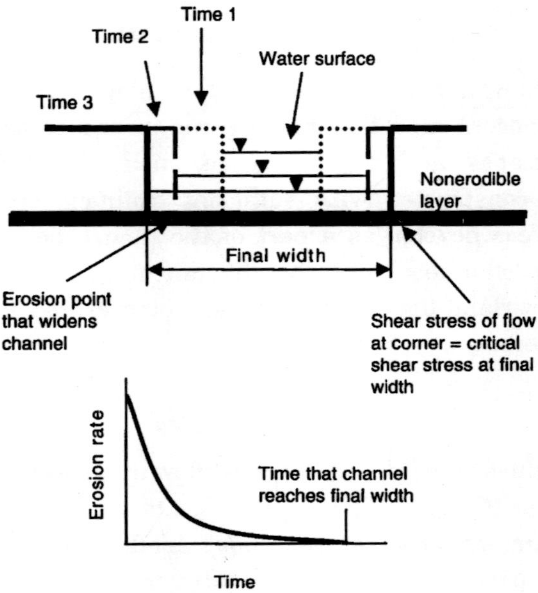

Figure 4.17 Effect of a nonerodible layer on erosion rate for different storm sequences.

the final width for that particular storm. Ephemeral gully erosion will be large if the difference in these widths is large; it will be small if the difference is small. If the final width is less than the existing width, no erosion occurs for the storm.

Interaction of Concentrated Flow Areas with Overland Flow Areas

Just as interrill areas deliver runoff and sediment to rill areas, overland flow areas deliver runoff and sediment to concentrated flow areas. An ephemeral gully begins at a finite point in a hollow on a converging landscape where overland flow merges into a single, definable channel (Figure 4.18). One definition for the beginning of an ephemeral gully is where the interrill slope length becomes zero because of the convergence of the rill areas on the converging landscape. In contrast to a terrace or diversion channel, an ephemeral gully has a drainage area and thus a flow rate and sediment load at the beginning (head) of the gully. Detachment can occur and transport capacity is greater than zero at the head of the gully. The flow rate increases along the gully as runoff from adjacent overland-flow areas enters the gully.

The concave channel profile of some ephemeral gullies means that the

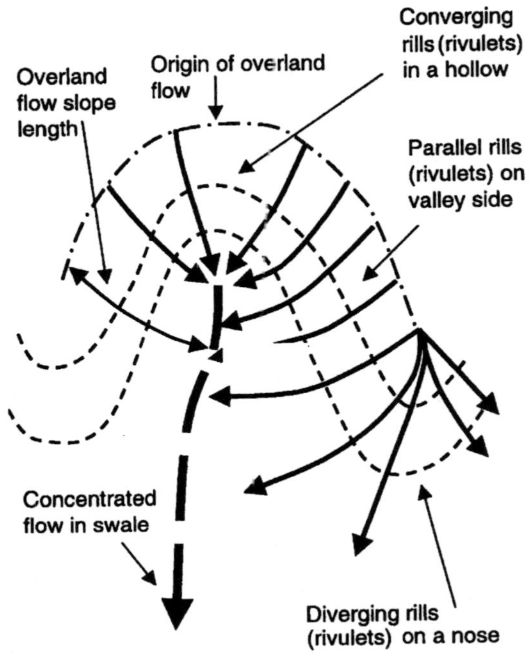

Figure 4.18 Definition of head of an ephemeral gully.

grade decreases continuously along the channel. Because the transport capacity is proportional to the product of flow rate and grade, the transport capacity first increases and then decreases along the gully (Figure 4.19). If the grade at the lower end of the gully is sufficiently flat, the transport capacity will decrease to less than the sediment load and deposition will occur on the lower reach of the gully. The sediment yield from the gully and its drainage area can be much less than sediment delivery from the overland-flow areas and from the upstream reach of an eroding ephemeral gully.

The runoff and sediment from many ephemeral gullies and similar concentrated-flow areas that occur within farm fields, construction sites, landfills, and reclaimed surface mines must flow through culverts, fence rows, and dense vegetated strips before reaching defined channels that connect directly with the stream-channel network. The limited flow capacity of culverts and high hydraulic roughness of fence rows and dense vegetated strips cause backwater that has a very low transport capacity, inducing much deposition.

A simple gradient terrace has a constant grade, which is usually suffi-

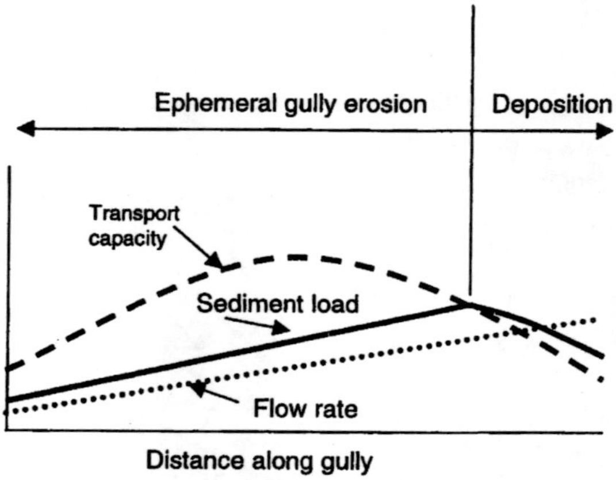

Figure 4.19 Erosion processes along an ephemeral gully with a concave profile.

ciently flat to cause deposition. Deposition occurs in a terrace channel when transport capacity in the terrace channel is less than the sediment load delivered to the terrace channel by the adjacent overland flow area. Deposition rate along the channel is uniform, and sediment load increases linearly along the channel. Figure 4.20 is a plot of deposition and sediment load along a simple terrace. Simple diversion channels are designed so as neither to cause deposition nor to allow erosion in the channel. The sediment yield from these channels is simply the amount of sediment delivered to them by the overland flow area.

Permanent, Incised Gullies

The processes in permanent, incised gullies are very much like those in rills that have not reached a nonerodible layer. Permanent, incised

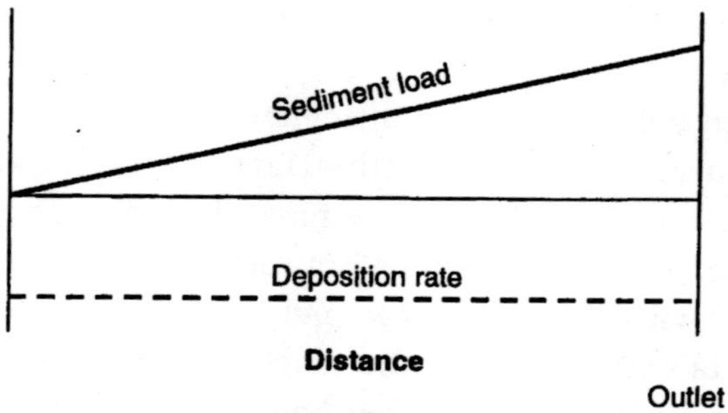

Figure 4.20 Deposition and sediment load along a simple, gradient terrace.

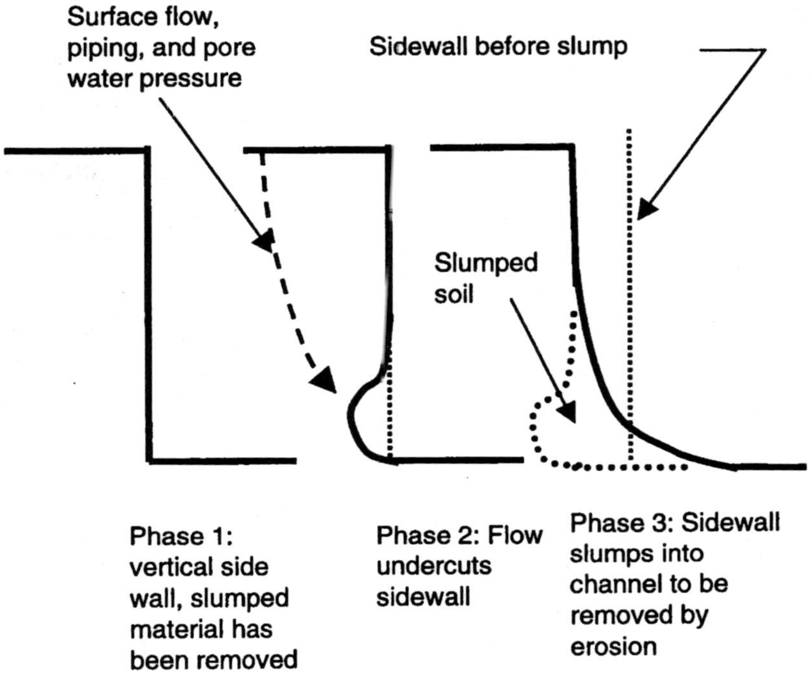

Figure 4.21 Phases in erosion of gully sidewall.

gullies, like those in Figure 3.8, form in concentrated flow areas by headcuts advancing upstream In addition, permanent, incised gullies are eroded by laterally retreating sidewalls where the process is cyclical, (Figure 4.21). The first phase is a vertical sidewall. The flow erodes and undercuts the sidewall. Pore-water pressure or piping from subsurface flow in the bank makes the soil more erodible, accelerating the rate at which the sidewall is undercut. Undercutting of the sidewall leaves a mass of hanging soil. As the undercutting proceeds, the weight of this soil exceeds the internal soil strength, and the soil "fails," resulting in it slumping into the channel. This slumped material is easily eroded, which significantly increases the local erosion rate and sediment load of the flow in the gully (Piest et al., 1975). After this increase, erosion rate and sediment load decrease as the material is cleaned out by erosion. Then the cycle repeats and continues to repeat until the gully becomes so wide that the erosivity of the flow is too small to undercut the gully sidewall, or vegetation develops on the slumped sidewall to stabilize it. The slumped material provides mechanical stability to the sidewall so that it does not slump further.

Erosion in Surface Irrigation Furrows

Erosion in furrows where water is surface applied for irrigation provides ideal study conditions for rill erosion processes and their interaction (Trout 1996). Flow is introduced at the upper end of the furrow (Figure 4.22). The flow rate decreases along the furrow because of infiltration.

The erosion rate is greatest at the upper end of the furrow, where the flow rate is highest and the sediment load is empty, and the flow erodes at its detachment capacity. The detachment rate decreases along the furrow because the detachment capacity decreases. Even though the detachment rate decreases along the furrow, sediment load accumulates. This filling of the transport capacity further decreases detachment rate.

One of three events occur, depending on the erodibility of the soil. If the soil is highly erodible, the sediment load accumulates rapidly along the furrow, while the transport capacity decreases because of the decreasing flow rate. The sediment load and transport capacity become equal. Thereafter, deposition begins and continues until the end of the furrow because the transport capacity continues to decrease.

Another possibility occurs when soil erodibility is low, which causes low erosion rates as well as causing the sediment load to accumulate slowly along the furrow. In this case, the location where deposition begins, which

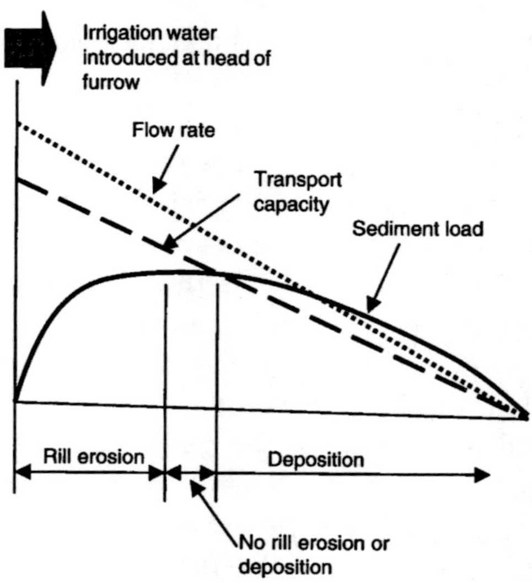

Figure 4.22 Erosion processes in surface irrigation furrow.

is the location where the sediment load equals the transport capacity, is much further along the furrow than for the highly erodible soil.

A third possibility occurs where the shear stress of the flow decreases to the soil's critical shear stress before the sediment load equals the transport capacity. The sediment load remains constant between the location where the flow's shear stress equals critical shear stress and the point where the sediment load equals the transport capacity because no erosion or deposition occurs in this region.

Stream Channels

Just as rill areas respond to interrill areas and concentrated-flow areas respond to overland-flow areas, stream-channels respond to their contributing watersheds. Stream-channel processes are described by Chien and Wan (1999), Chow (1959), Knighton (1984), Richards (1981), and Schumm (1977).

Small Impoundments

Naturally occurring and constructed impoundments dramatically reduce sediment movement from the landscape. Naturally occurring impoundment include the potholes left by glaciers. Small constructed impoundments include those used in association with terraces on farm fields where a small embankment is placed across a concentrated-flow area. Another example is the small sediment-control basins used on construction sites to control sediment yield during construction operations. Other common small impoundments consists of backwater caused by culverts, roadways, fencerows, and grass strips at the edge of fields.

Generally, large amounts of deposition occur in impoundments because the very slow flow through them has essentially no transport capacity. The amount of deposition is controlled by retaining water in the impoundment for a sufficient duration that the sediment settles to the bottom of the impoundment before the flow leaves the basin (Haan et al., 1994). A major determinant in the sediment-trapping efficiency of reservoirs is the size and density of the sediment particles reaching the impoundment. If the incoming sediment particles are very coarse and dense, almost all of the sediment will be trapped because these particles settle rapidly. In contrast, the trapping efficiency of very fine sediment is much lower than that of coarse sediment because fine sediment settles very slowly. A longer flow retention time in the impoundment is needed to trap fine sediment than is needed to trap coarse sediment. In general, sediment-control basins are highly effective at trapping sediment, leaving only the very fine particles in the flow discharged from the impoundment. The second impoundment in a series of two impoundments or an impoundment downstream of a well-performing vegetative strip may not trap a high percentage of the incoming sediment load because upstream deposition has already removed the coarse particles and the remaining fine particles are not easily deposited. Also, reservoirs lose their sediment-trapping efficiency as they fill with sediment.

Relation of Sediment Yield to Upland Erosion

The mass-balance equation that describes the variables that define sediment yield in relation to upland erosion and other erosion variables is

$$S_y = A_{ir} + A_{ch} - D_{up} - D_{val} \qquad (4.1)$$

where S_y is the sediment yield, A_{ir} the interrill + rill erosion on the eroding portions of the landscape where overland flow occurs, A_{ch} the concentrated-flow erosion, D_{up} the deposition on upland areas, including the overland-flow portion of the landscape and areas of concentrated flow, and D_{val} the sedimentation on valley floors. Sediment yield is frequently estimated by the equation

$$S_y = \mathrm{SDR} \cdot A_{ir} \qquad (4.2)$$

where SDR is the sediment delivery ratio. In general, values for SDR are assumed to decrease with watershed size (Haan et al., 1994). For example, the SDR for a 65-acre (25-ha) watershed is about 0.5, whereas it is 0.1 for a watershed as large as 65,000 acres (25,000 ha). Sediment delivery ratios are commonly assumed to be less than 1 because of deposition in the floodplains. However, deposition occurs in many other locations of a watershed, including on concave slopes. As much as 75% of the soil loss eroded on a complex slope can be deposited on the lower concave portion, never reaching a concentrated flow even within a field-sized area (Foster et al., 1980; Lu et al., 1988; Piest et al., 1977). Dense grass strips as narrow as 18 in. (0.5 m) can trap up to 90% of the sediment reaching them (Flanagan et al., 1989). More than 50% of the sediment eroded in an agricultural field can be deposited within the field by overland flow on concave hillslopes and in the backwater of concentrated-flow areas (Foster, 1985). Thus, a large portion of the sediment produced on the eroding portions of hillslopes may be deposited nearby the source area, never to reach a stream channel. The degree to which such deposition occurs depends strongly on hillslope shape. Although concave slope shapes can induce large amounts of deposition, no deposition occurs on convex hillslopes.

Although many stream channels and the floodplains alongside them experience deposition, some streams can be unstable and produce most of the sediment being transported in the stream. For example, stream-channel erosion for watersheds in the Bluffline Region of northwestern Mississippi accounts for about 80% of the sediment load in those channels (Kuhnle et al., 1996).

Equation (4.1) can be rearranged to show the variables that influence SDR as

$$\text{SDR} = \frac{S_y}{A_{\text{ir}}} \tag{4.3}$$

or

$$\text{SDR} = \frac{A_{\text{ir}} + A_{\text{ch}} - D_{\text{up}} - D_{\text{val}}}{A_{\text{ir}}} \tag{4.4}$$

which shows that SDR is much more complicated that simply being a function of watershed size. A common assumption is that SDR does not change with a change in land use. The fundamentals associated with whether sediment available or transport capacity controls sediment delivery shows that SDR is not a constant value as land use changes.

In some cases, upland erosion is estimated from sediment yield values as

$$A_{\text{ir}} = S_y - A_{\text{ch}} + D_{\text{val}} + D_{\text{up}} \tag{4.5}$$

Sediment yield, S_y, can be measured easily in stream channels, but measuring the other variables, especially deposition on upland areas, D_{up}, is almost impossible. Therefore, accurately estimating upland erosion from sediment yield values is very inaccurate because of the difficulties in measuring the other terms in the sediment mass balance equation.

WIND EROSION

This detailed description of wind-erosion processes is organized around the processes themselves rather than the spatial context used to describe water erosion. A major difference between water and wind erosion is that while runoff and sediment always flow in the downslope direction, wind can come from any direction during any particular storm, although wind directions follow a general pattern according to location and season. This discussion of wind erosion begins with the detachment processes by drag, lift, and abrasion forces, followed by the transport modes of creep, saltation, and suspension.

Basic Mechanics

Wind erosion and sedimentation results from the combined effects of detachment, entrainment, transport, and deposition of soil particles. Wind across a soil surface exerts drag and lift forces that detach soil particles, entrains previously detached (loose) particles, and transports sediment. Windblown sediment is also an erosive agent itself that detaches sediment by abrasion. Deposition occurs when and where the sediment load is greater than the wind's transport capacity.

Detachment by Wind Forces

Erosivity of the Wind

To detach sediment particles, the wind exerts both drag and lift forces to overcome the gravitational and cohesive forces that hold soil particles in place on the soil surface. Detachment requires that wind forces exceed the soil's resisting forces.

Both drag and lift forces are proportional to the square of the wind velocity, which means that these forces increase much more rapidly than does wind velocity. For example, doubling of wind velocity quadruples drag and lift forces. The drag force applied to the soil can be determined from the velocity gradient of the wind at the soil surface (Figure 4.23).

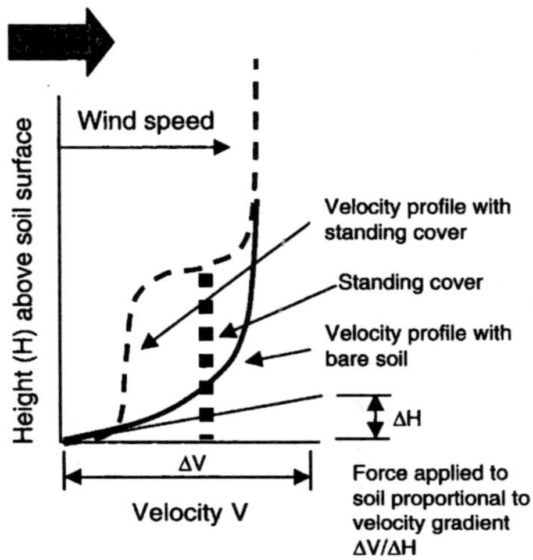

Figure 4.23 Wind velocity profile over a soil surface with bare and standing cover.

A drag force (shear stress) detaches sediment particles from the soil in the same way that sand particles can be rubbed from sandpaper. Imagine a piece of sandpaper laying on a table with the grain-side upward and rubbing your hand across the rough surface of the sandpaper. A glue bonds the sand particles to the paper backing. The bonding force of the glue is analogous to the cohesive bonding forces that hold soil particles in place. The sand particles can be rubbed from poor-quality sandpaper quite easily, whereas substantial force is required to rub the sand particles from high-quality sandpaper. Just as sandpaper varies in the bonding strength of the glue, soils vary in their internal bonding strengths among soil particles. The wind's lift force is the same as the lift force created on an airplane wing. Lift forces during a windstorm elevate soil particles from the soil surface.

Wind velocity at the soil surface is zero and increases with height on a smooth, bare soil surface.

Vegetative cover and rough soil reduce the erosivity of wind by reducing the forces that the wind applies to the soil surface. Figure 4.23 shows how standing cover, flat-lying cover, roughness, and ridges generally affect the wind's velocity profile in relation to the wind-velocity profile for a smooth, bare soil. These surface conditions reduce the wind's velocity gradient at the soil surface, which means that the forces applied by the wind to the soil surface are reduced similarly. Flat cover is nonerodible material, such as gravel and crop residue, laying on the soil surface. The thickness (height) of this cover is negligible. Flat cover reduces the velocity gradient at the soil surface but does not change the basic shape of the velocity profile. The effect of flat cover on erosion is related to the percent of the soil surface covered by the material and the effect of flat cover is independent of wind direction if the cover is uniform. If flat cover is in strips, however, the effect of flat cover depends on the orientation of the strips with respect to wind direction. Strips of flat cover most effectively reduce wind erosivity when perpendicular to the wind direction and are least effective when parallel with the wind direction (Fryrear and Skidmore, 1985).

Standing cover is erect material, such as live plants or stubble after harvest of a crop, that projects into the wind. This material reduces the velocity gradient at the soil surface and changes the shape of the velocity gradient within and immediately above the plants. Standing cover reduces wind erosivity more than the same mass of flat cover (Fryrear and Skidmore, 1985). The effect of standing cover depends on the height and areal and mass density of the cover, open space within and between the plants, and the gradient of open space with height (Bilbro and Fryrear, 1994). The gradient of open space refers to how the percent of open space per unit height changes with height above the soil surface. Plants having most of their open space near their top reduce wind erosivity less than do plants having most of their open space near their bottom. If the cover is uniform in the plan view, the effect of standing cover is independent of wind direction. However, if standing cover is in strips or rows, which is typical of agricultural crops such as corn, the effect of standing cover is a function of the orientation of the strips or rows with respect to wind direction.

Barriers, narrow strips of dense vegetation including tall grass and trees, are placed around erodible areas to reduce the wind erosivity in the vicinity of the barriers (Chapters 2 and 7; Fryrear and Skidmore, 1985). The effect of the barriers extends both upwind and downwind from the barrier, although the effect extends farther downwind than upwind (Figure 4.24).

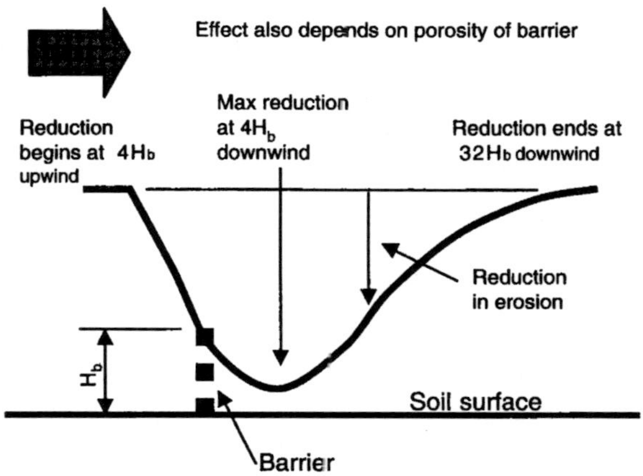

Figure 4.24 Effect of barrier on wind velocity.

Both random and oriented roughness (ridges) affect wind erosivity. The effect of random roughness is independent of wind direction, but the effect of ridges depends on the orientation of the ridges with respect to wind direction. The overall effect of random roughness and ridges perpendicular to the wind is to reduce the wind-velocity gradient at the soil surface. However, the effect of roughness and ridges on erosivity is best examined on a micro (local) scale. Although the net effect is an overall reduction in erosivity, the forces applied by the wind to random roughness and ridges are increased on the peaks of the roughness elements and ridge tops, and decreased in the depressions and furrows. This distribution of local erosivity affects local detachment and deposition and selective erosion of particle classes.

Soil Resistance to Detachment by Wind Forces

The other major factor in addition to wind erosivity that determines particle detachment by wind is soil erodibility. Two types of forces determine soil erodibility: (1) gravity that holds soil particles in place (weight of the soil particles), and (2) the internal cohesive bonding forces among soil particles. A noncohesive soil such as sand has only gravitational forces holding the particles in place. The fundamental soil conditions that determine soil resistance to detachment include texture, organic-matter content, crusting, soil moisture, and roots (Chapter 2; Chepil, 1956; Woodruff and Siddoway, 1965; Zobeck, 1991).

Mechanical soil disturbance that roughens the soil, creating large clods (large soil aggregates), reduces wind erosion. The size of these clods is so large that the drag and lift forces of the wind cannot move these particles. In effect, these clods armor the soil surface to reduce detachment significantly. The effect of these aggregates is similar to that of stones that are too large for the wind to entrain that armor the soil surface. The critical size of nonerodible stones is less than the critical size of soil aggregates because the density of the stones is much greater (specific gravity of 2.65) than the density of the aggregates (specific gravity of about 1.4).

Another benefit of a rough surface is that the easily eroded, fine soil aggregates occupying the depressions are protected in areas where the local wind forces are low. Random roughness can be created by mechanically shaping small aggregates into large random-roughness shapes. However, the fine aggregates are not resistant to wind forces on the roughness peaks. The peaks will be eroded rapidly and the depressions filled with sediment. Thus, random roughness is most effective in reducing wind erosion when formed with large, erosion-resistant clods.

Detachment by Abrasion

Erosivity of Abrading Sediment

Abrasion is the other important detachment mechanism of wind erosion. Windblown sediment particles striking the soil surface (Figure 4.25) function as miniature ballistic projectiles that detach soil particles (Lyles et al., 1985). The amount of sediment detached by abrasion can be several times the total sediment detached during a windstorm by shear forces (Lyles et al., 1985).

A sediment particle striking the soil surface exerts instantaneous

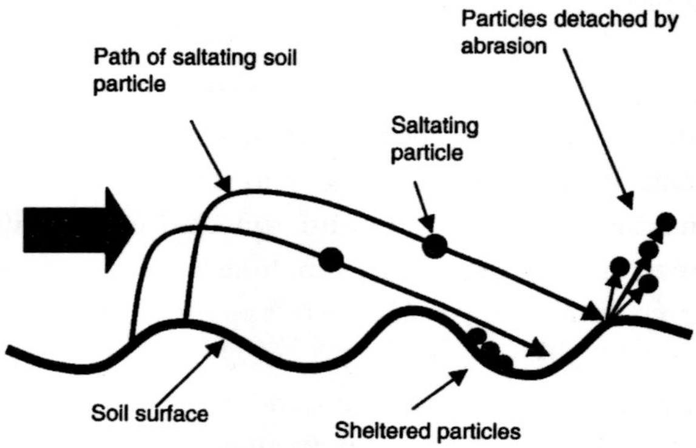

Figure 4.25 Soil detachment by abrasion.

forces on the soil. Impact energy is used as a measure of the erosivity by abrading particles. The impact energy of an individual particle is directly proportional to the mass of the particle and the square of its impact velocity. The mass of a sediment particle is a function of the diameter of the particle and the specific gravity of the particle. The impact energy of a sand grain with a specific gravity of 2.65 is nearly twice the impact energy of a soil aggregate that has a specific gravity of 1.3. The total erosivity of the abrading sediment is proportional to the total mass of sediment particles that strike the soil. The impact energy of a 0.5-mm-diameter particle is 125 times the impact energy of a 0.1-mm particle at the same impact velocity.

Abrading sediment is transported primarily by saltation. Sediment transported by creep detaches little sediment by abrasion because the impact velocity of these particles is very low. Sediment transported by suspension detaches little sediment by abrasion because of low impact energy of the small mass of the individual particles and because little of the suspended sediment comes into contact with the soil surface.

For abrasion to occur, sediment must first be detached by the wind's drag and lift forces. The process of abrasion grows exponentially as the saltating sediment load accumulates. Thus, cover, roughness, and the other factors that reduce detachment and transport capacity for the saltating sediment load also reduce the erosivity of abrading sediment and detachment by abrasion.

Resistance to Abrasion

The crushing energy of aggregates is a measure of a soil's strength and resistance to abrasion (Hagen et al., 1992). The crushing energy is measured by placing an aggregate in a testing device that applies a compressive force and determines the energy required to fracture the aggregate. Just as a crusted soil is highly resistant to detachment by the wind's drag and lift forces, a crusted soil is highly resistant to abrasion. The factors that increase the soil's resistance to shear forces also increase the resistance of a soil to abrasion. In addition, random roughness and ridges provide a sheltering effect. Abrading sediment strikes the soil surface at an angle (Figure 4.25), and the soil on the back side of a roughness element or a ridge is sheltered. Conversely, soil particles on the front side of a roughness element or ridge are exposed to greater abrasive force than is soil on a smooth surface.

A characteristic size and density distribution of sediment is produced when abrasion detaches sediment particles. In addition, the impact itself can fracture and reduce the size of the windblown sediment doing the abrading. Thus, the average size distribution of the sediment is finer when much of the sediment is detached by abrasion than when little sediment is detached by abrasion.

Entrainment and Transport by Wind

When sediment is detached initially, it is entrained by the wind and transported at least a short distance. A detached particle, however, may move only a short distance before coming to rest. Entrainment is the process of initiating movement of a previously detached or noncohesive particle. The wind must overcome only gravitational forces to entrain and transport these particles.

Transport occurs by the three modes of creep, saltation, and suspension (Figure 4.26). When sediment is transported as creep, the sediment moves along the surface by rolling or sliding. The factors that determine sediment-transport capacity for creep are the shear stress applied to the soil surface by the wind and the size and density of the sediment particles. Drag forces at the soil surface are more important than lift forces in the transport of particles by creep.

When sediment is transported as saltation, the sediment particles are lifted from the soil surface and follow a trajectory illustrated in Figure 4.26. The lift forces applied to the sediment particles initiate particle

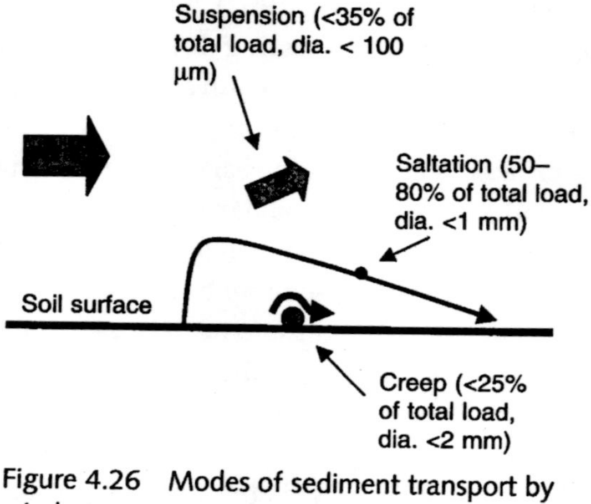

Figure 4.26 Modes of sediment transport by wind.

movement like a gun charge initiates a projectile's flight. As the particle rises into the wind, the wind's drag force carries the particle downwind and the particle gains horizontal velocity. At the same time, the particle loses velocity in the vertical direction, reaches a maximum height, and descends back to the soil surface. The sediment particle strikes the soil surface with both vertical and horizontal velocity. The sediment-transport rate by saltation is a function of the size and density of the particles, the frequency with which the particles are launched, and the average particle velocity over the saltating path (Yalin, 1963).

The maximum height that a sediment particle reaches on a saltating trajectory depends on the size and density of the particle and the wind forces. When the particle's size and density are sufficiently small, wind forces may carry the sediment sufficiently high that it becomes a part of the suspended load. Wind turbulence keeps particles suspended, resulting in the transport of suspended particles for long distances before being deposited, especially for fine, dust-sized particles. A sediment-concentration gradient develops with height in the wind, such that the concentration is greatest at the soil surface and decreases upward. A similar size and density gradient develops as well, with the largest and densest particles in the suspended load traveling nearest the soil surface (Stout and Zobeck, 1996). Suspended sediment moves higher in the wind with distance downwind (Figure 4.27).

Each mode of transport has its own transport capacity. Standing and flat cover, random roughness, and ridges oriented perpendicular to the wind direction reduce the wind forces at the soil surface and thus reduce the wind's transport capacity for each mode. These surface conditions affect each mode of transport differently. For example, standing cover causes a major reduction in the transport capacity for creep and saltation but causes much less reduction in the transport capacity for suspended sediment, especially where the suspended sediment concentration gradient is well developed. If the suspended sediment is sufficiently high in the wind, transport capacity for the suspended sediment is unaffected by local cover or roughness.

Deposition of Windblown Sediment

The sediment transported by wind is deposited when and where the sediment load for a particular mode of transport exceeds that mode's transport capacity (Figure 3.11). Deposition is a selective process such that the

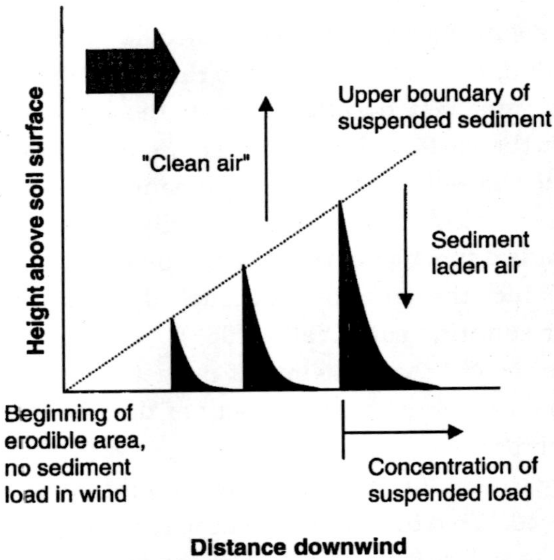

Figure 4.27 Development of sediment load by suspension.

largest and densest particles are deposited first and the smaller and less dense particles are transported farther downwind before being deposited (Figure 4.2).

Interaction of Detachment, Sediment Transport, and Deposition

Sediment detachment, transport, and deposition interact on both the micro and macroscales. On the microscale, interaction occurs with both random and oriented roughness (ridges). The forces are greatest on the peaks of the ridges and least in the furrows. Detachment on the ridges produces a local sediment load that exceeds the local transport capacity in the furrows, causing deposition in the furrows. Detachment on the ridges reduces ridge height, and deposition in the furrows fills them, reducing the effect of ridges through time. As the ridges are eroded and the furrows are filled, the differential in detachment and deposition decreases, so that local deposition ends as the soil surface becomes smooth. This local deposition enriches the sediment load in fine particles and the soil surface in coarse particles, illustrating one way that wind erosion is selective.

On the macroscale, consider the field illustrated in Figure 4.28. The sediment load is assumed to be zero at the upwind edge of the field, and

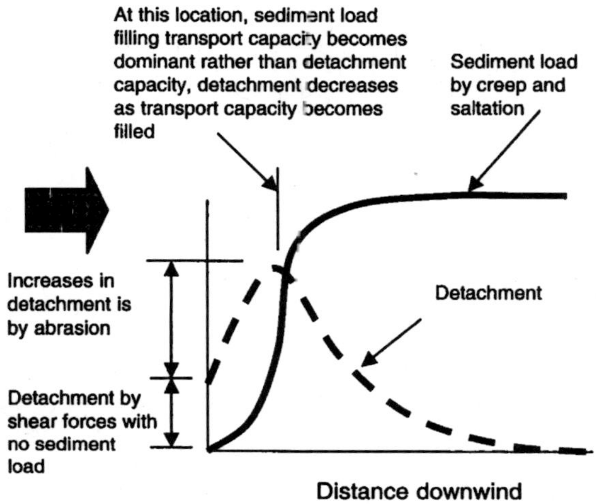

Figure 4.28 Relation of detachment and sediment load.

the erosivity along the field is assumed to be uniform. The detachment rate from shear forces is significant at the beginning of the field and continues to increase as a result of abrasion with an increase in sediment load. At a point in the field, however, the detachment rate begins to decrease and continues to decrease with distance across the field (Stout, 1990; Stout and Zobeck, 1996). This decrease occurs as the sediment load approaches the transport capacity of the wind for the medium-sized and coarse sediment transported by creep and saltation (Stout, 1990; Stout and Zobeck, 1996). The transport capacity for the suspended load generally is not limiting, and the sediment load of the fine particles continues to increase along the field. Whereas the transport capacity can limit the sediment load of medium-sized and coarse particles in high-erosion-rate situations, the sediment load of fine particles more often is limited by detachment, illustrating the concept of detachment- or transport-limiting erosion conditions.

Deposition occurs where the transport capacity is less than the sediment load. An example is the fence row in Figure 3.11 that reduces transport capacity. The rate of deposition is proportional to the differences in transport capacity, sediment load, fall velocity of the sediment, wind speed, and height of the sediment in the wind. Deposition rates are high where the difference between transport capacity and sediment load is large. The deposition rate also is greatest for those particles that have a high fall velocity in still air, which are the largest and densest particles.

These particles are deposited just downwind of the location where the transport capacity decreased (Figures 3.11, 4.1, and 4.2). The smaller and less dense particles are transported farther downwind before being deposited because of their low fall velocity.

Wind-erosion selectively removes the fine particles from a soil and leaves the coarse particles (Lyles and Tatarko, 1986). This selectivity is caused exclusively by entrainment and deposition and not by detachment. The differential rates of deposition illustrated in Figure 4.2 is one example of selective wind erosion, leaving coarse particles and transporting fine particles downwind. Another example is deposition in roughness elements formed by random roughness or by ridges oriented perpendicular to the wind.

Although wind erosivity is assumed to be uniform over a sizeable area, based on average wind velocity. The erosivity of the wind on a local scale of less than 3 ft (1 m) is very nonuniform, which means that the local detachment and transport capacity of the wind is also very nonuniform. When a soil surface is highly erodible and the wind is very erosive, the sediment load rapidly fills to the transport capacity, such that the local transport capacity can decrease below the local sediment load, especially for sediment transported by creep. This coarse sediment is deposited in depressions formed by random roughness or in furrows between the ridges oriented perpendicular to the wind direction. This deposition enriches the soil surface in coarse particles and enriches the sediment load in fine particles. Through time, the soil becomes increasingly coarse.

Detachment of sediment from cohesive soils is nonselective. Small particles cannot be removed from cohesive soil without breaking the cohesive bonds among soil particles. When these bonds are broken, all particle classes are released into the wind. However, abrasion can produce particles larger than the wind can transport, creating loose sediment on the soil surface. Entrainment can be selective and remove the small particles, leaving the coarse particles. This selective process cannot continue indefinitely. Eventually, coarse particles such as gravel or nonerodible aggregates accumulate on the soil surface, which armor and protect the soil surface from further erosion. Where erosion was first limited by transport capacity, now erosion becomes limited by detachment as the surface becomes armored with coarse particles.

SUMMARY

Erosion and sedimentation involve the processes of detachment, entrainment, transport, and deposition of soil particles by the erosive agents wind and water. Sediment is detached when forces applied to the surface soil properties exceed the resistance of the soil to these forces. Particles are entrained and transported when the lift and drag forces applied exceed the weight and frictional force holding the particles in place. Sediment load is limited by the amount or sediment made available for transport by detachment or by the transport capacity of the flow. Deposition occurs when and where sediment load exceeds transport capacity. Deposition is proportional to the difference between transport capacity and sediment load, sediment fall velocity, height of the sediment within the flow, and flow velocity. Although detachment is nonselective, entrainment and especially deposition are highly selective, resulting in coarse particles being deposited and fine particles being transported farther downstream before deposition. Selective entrainment eventually results in an armored surface of coarse particles.

SUGGESTED READINGS

Foster, G. R. 1985. *Understanding ephemeral gully Erosion (Concentrated Flow Erosion)*. In: Soil Conservation, Assessing the National Resources Inventory. National Academy Press, Washington, D.C.

Foster, G. R., R. A. Young, M. J. M. Römkens, and C. A. Onstad. 1985c. *Processes of soil erosion by water*. In: Soil Erosion and Crop Productivity, R. F. Follett and B. A. Stewart, (eds). Amer. Soc. Agron. Crop Sci. Soc. of Amer., Soil Sci. Soc. of Amer. Madison, WI.

Fryrear, D. W. and E. L. Skidmore. 1985. *Methods for Controlling Wind erosion*. In: Soil Erosion and Crop Productivity, R. F. Follett and B. A. Stewart, (eds). Amer. Soc. Agron. Crop Sci. Soc. of Amer., Soil Sci. Soc. of Amer. Madison, WI.

Lyles, L., G. W. Cole, and L. J. Hagen. 1985. *Wind erosion: Processes and Production*. In: Soil Erosion and Crop Productivity, R. F. Follett and B. A. Stewart, (eds). Amer. Soc. Agron. Crop Sci. Soc. of Amer., Soil Sci. Soc. of Amer. Madison, WI.

5

Erosion-Prediction Technology

Erosion-prediction technology is a powerful tool used for more than half a century in policy development, erosion inventories, conservation planning, and engineering design (Chapter 8). Erosion-prediction technology consists of mathematical equations that compute values for erosion variables using input values for climate, soil, topography, and land use (Chapter 2). A particular set of these equations is known as a *mathematical erosion model*. Various erosion models compute estimates of soil loss, deposition, sediment yield, and characteristics of the sediment being transported. Many erosion models are available, each with its particular features, resource requirements, ease of use, strengths, and limitations. The computer program for a specific model, databases used as input to the model, and supporting documentation including a user guide can often be downloaded from an Internet site for the model (Appendix C). In this chapter we describe the fundamentals of erosion-prediction technology, elements of mathematical equations used in erosion models, types of erosion models, steps in developing an erosion model, choosing an erosion model, and conducting a sensitivity analyses for an erosion model.

FUNDAMENTALS OF EROSION-PREDICTION TECHNOLOGY

Every erosion model must represent how the four factors of climate, soil, topography, and land use affect soil loss and related variables. A simple erosion model is

$$\mathrm{SL} = \mathrm{CF} \cdot \mathrm{SF} \cdot \mathrm{TF} \cdot \mathrm{LUF} \tag{5.1}$$

where SL is the average annual soil loss, CF the climate factor, SF the soil factor, TF the topographic factor, and LUF the land-use factor. A value for average annual soil loss is computed for a particular field site with a specific land use by substituting values for each factor in the equation. An example computation is

$$\begin{aligned} \text{SL} &= 200 \cdot 0.32 \cdot 1.2 \cdot 0.2 \\ &= 15 \text{ tons/acre (33 metric tons/ha) per year} \end{aligned} \tag{5.2}$$

where 200 erosivity units per year represents the climate-factor value based on location of the site, 0.32 is the mass/area per erosivity unit, the soil-factor value based on the soil properties at the site; 1.2 is the topographic-factor value for the field length or slope length and steepness of the site, and 0.2 is the land-use factor value based on existing land use at the site. If the soil-loss tolerance value (T) is 5 tons/acre (11 metric tons/ha) per year for this soil (Chapter 8), soil loss is too great with the current land-use practice. Another land-use practice is needed that reduces estimated soil loss to soil-loss tolerance. The maximum allowable value for the land-use factor (MALUF) is computed as

$$\text{MALUF} = \frac{T}{\text{CF} \cdot \text{SF} \cdot \text{TF}} \tag{5.3}$$

$$\begin{aligned} &= \frac{5}{200 \cdot 0.32 \cdot 1.2} \\ &= 0.065 \end{aligned} \tag{5.4}$$

Any land-use practice having a land-use factor value below 0.065 provides acceptable erosion control for the site.

ELEMENTS OF EROSION-MODEL MATHEMATICS

The mathematical equations used in erosion models have five components: (1) independent variables, (2) dependent variables, (3) parameters, (4) mathematical operators, and (5) a computation sequence and logic that links the equations within the model. The independent variables represent the basic factors, including erosivity, soil erodibility, topography, canopy cover, ground cover, surface roughness, and below ground biomass, which collectively determine erosion (Chapters 2 and 4). Values for the independent variables are determined from the input information entered in the model. The dependent (output) variables include soil loss and may include deposition, sediment load, and sediment characteristics, depending on the model's capabilities.

Parameters are terms in the mathematical equations that modify values for the independent variables as a part of the mathematical computations. For example, the area of a square A_s is given by

$$A_s = x^2 \tag{5.5}$$

where x is the length of the side of the square. The exponent in equation (5.5) is a parameter, and 2 is the value of the parameter. The length of the side, x, is the independent variable (input variable), and the area, A_s, is the dependent variable (output variable). The effect of ground cover in some erosion models is computed using

$$\text{GCF} = \exp(-b \cdot \text{percent ground cover}) \tag{5.6}$$

where GCF is the ground cover factor and b is a parameter whose value is a function of the ratio of rill to interrill erosion with bare soil conditions. Parameter values for an equation are determined by fitting an equation such as equation (5.6) to experimental data or by deriving an equation from theory.

Equations involve such operations as addition, multiplication, and raising numbers to powers. The exp in equation (5.6) is mathematical notation for raising the natural logarithm base e of 2.718 to the power of ($-b \cdot$ percent ground cover). The perimeter of a rectangle, P_r, is given by

$$P_r = (2 \cdot \text{width}) + (2 \cdot \text{length}) \tag{5.7}$$

The operators are the two multiplications and the single addition. The parentheses are used to group the variables and mathematical operators to ensure that the operations are performed in the desired sequence.

Not only must the mathematical operations occur in the proper sequence within an equation, but the equations within an erosion model must be executed in the proper sequence. For example, equation (5.6) cannot be executed in the model until equations have been executed that determine values for the parameter b and the independent variable of ground cover. Choices between equations are a typical part of an erosion model. For example, deposition occurs when the sediment load exceeds transport capacity (Chapter 4), which requires a logic check to compare sediment load with transport capacity. This particular example would be executed as

$$\text{IF sediment load} > \text{transport capacity} \quad \begin{array}{l}\text{THEN compute deposition}\\ \text{ELSE compute detachment}\end{array} \tag{5.8}$$

where IF is a logic operator, THEN is the action when the IF check is true, and ELSE is the action taken when the IF check is false.

Both temporal and spatial variability must be considered. The independent variables controlling erosion vary over time and space, so the dependent erosion variables also vary over time and space. The simplest erosion models compute soil loss for a slope length without regard to how detachment and deposition vary along the slope length. These models also use average annual values for inputs without consideration of temporal variations during the year. The most complex erosion models compute values at discrete points in time and space, including time between storms and within a storm. These point values in time and space are integrated numerically to determine soil loss for slope length, field length, field, watershed, or region. Also, the point values show where erosion is most severe and the time during a crop rotation or the recovery period after a land disturbance, such as mine reclamation, logging, or construction, when the erosion rate is highest.

The main governing equation in all erosion models is the conservation-of-mass equation. Although a model may not explicitly solve the conservation of mass equation, the model must obey conservation-of-mass principles. The conservation-of-mass equation involves both time and space and is very difficult to solve in its complete form. The equation can be simplified by assuming steady state and solved to produce soil-loss equations that are used in erosion models. The steady-state conservation of mass equation is given by

$$\frac{\Delta G}{\Delta x} = D_{\text{net}} \tag{5.9}$$

where $\Delta G/\Delta x$ is the change in sediment load G with distance x and D_{net} is the net detachment or deposition rate. Equation (5.9) applies to a point

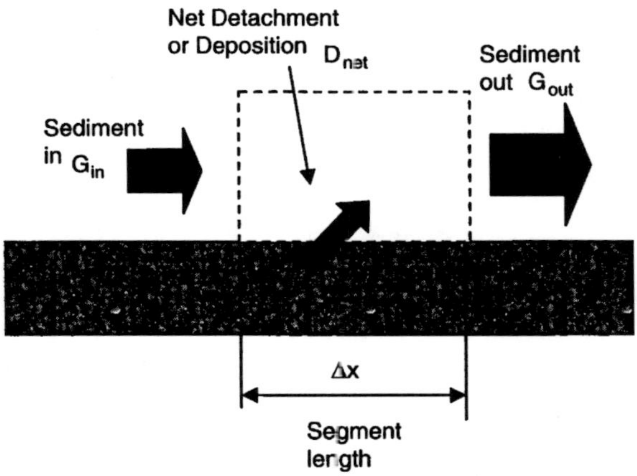

Figure 5.1 Schematic of segment on slope used to solve conservation of mass equation.

in space. Sediment load G is the variable of interest, and it can be determined by solving equation (5.9) as

$$G_{out} = G_{in} + \Delta x \cdot D_{net} \tag{5.10}$$

where G_{out} is the sediment load leaving the segment illustrated in Figure 5.1, G_{in} the sediment load coming into the segment, and Δx the length of the segment. Values for sediment load can be determined along the flow path (over space) by starting at the upstream (upwind) end of the flow path and "stepping through space" along the flow path by solving equation (5.10) segment by segment. An equation based on variables such as rainfall intensity, runoff rate, wind velocity, soil, slope steepness, and cover conditions is used to compute net detachment or deposition for each segment (see Chapters 2 and 4). Table 5.1 illustrates this numerical computation for a complex slope where the values for net detachment/deposition are assumed to be known from equations for these variables. A sign convention for detachment and deposition is needed. In this model, a positive sign is used for detachment because detachment increases the sediment load, and a negative sign is used for deposition because deposition decreases the sediment load.

The conservation of mass equation that involves time is illustrated in Appendix B. Fundamentally, the equation is given by

$$\frac{\Delta S_{tor}}{\Delta t} = (G_{in} + \Delta x \cdot D_{net}) - G_{out} \tag{5.11}$$

Table 5.1 Sediment Load Computations along a Complex Slope

Segment	*Distance to Lower End of Segment,* x_e	*Net Detachment / Deposition Rate,*[a] D_{net}	*Sediment Produced on Segment,* $\Delta x \cdot D_{net}$	*Sediment Load at Lower End of Segment,* G_{out}
1	15	3.1	46	46
2	30	10	150	196
3	45	22	330	526
4	60	39	585	1112
5	75	52	780	1892
6	90	41	615	2506
7	105	25	375	2882
8	120	−51	−765	2116
9	135	−31	−465	1651
10	150	−35	−525	1126

[a]Values for net detachment/deposition are assumed to be known from other computations.

where $\Delta S_{tor}/\Delta t$ is the storage rate (i.e., the change in storage of sediment in the spatial segment over time interval Δt), $G_{in} + \Delta x \cdot D_{net}$ the rate that sediment enters the segment during time interval Δt from the segment immediately upstream (upwind) and the gain or loss of sediment by net detachment or deposition within the segment, and G_{out} the rate that sediment leaves the segment during time intervalΔt. In simple terms, the storage rate in the segment equals the rate of inflow less the rate of outflow. If the sediment inflow rate, $G_{in} + \Delta x \cdot D_{net}$, is greater than the outflow rate, G_{out}, sediment accumulates within the segment, and conversely, sediment diminishes if the outflow rate exceeds the inflow rate. Equation (5.11) is solved numerically by stepping through time and space, just as equation (5.10) is solved by stepping through space.

The power of modern personal computers dramatically expands the equations that can be solved in an erosion model. However, complex erosion models typically require collection of extensive and sometimes difficult-to-obtain input data. The utility of the output from a complex model does not always increase in proportion to the increased costs of using the model. Some users prefer simple, easy-to-use, although less powerful erosion models. Types of models based on model structure are summarized in Table 5.2. In the following section we describe these major types of erosion model and some of their distinguishing features.

Table 5.2 Summary of Types of Erosion Models

Model Type	*Form*	*Derivation Method*	*Strengths*
Regression-derived	A single or a few equations having a form that best fits the data	Derived by fitting an equation(s) to an empirical database representing field conditions	Generally simple and easy to use: input values can be simple and easy to obtain
Index-based	Uses indices, usually in a multiplicative form, to represent how climate, soil, topography, and land use affect erosion	Values for indices determined from large empirical database representing field conditions	Simple and easy to use; input values can be simple and easy to obtain; very powerful in relation to simplicity and input values
Simple process-based	Represents individual erosion processes using simple steady-state equations	Equations derived from theory and empirical databases for erosion processes, validated against database representative of field conditions	Can be simple; represents main fundamental erosion processes; improved performance
Combined index and process-based	An index-based form in which simple process-based equations are used to determine values for indices	Components derived from empirical database for erosion processes and land-use subfactors and theory, validated against database representative of field conditions	Captures the best of both index-and simple process-based models; robust; uses simple, easy-to-obtain inputs; land-use independent, very powerful considering easy-to-obtain input values
Dynamic between storms	Uses equations to represent how variables controlling erosion change between storms, uses representative storm values and a steady-state equation to compute the erosion for each storm; form of equations typically based on theory	Derived from theory and empirical database for fundamental erosion processes, equations connected by conservation of mass equation, validated using database representative of field conditions	Powerful; can represent a wide range of conditions, captures main effects of fundamental erosion processes without excessive detail, can be constructed to be land-use independent and to use relatively simple inputs
Dynamic within and between storms	Has the additional feature of computing erosion through time within the event	Same as other dynamic model but with additional databases required for temporal variations within storm event	Most powerful of all erosion models, represents erosion processes to maximum detail

TYPES OF MATHEMATICAL EROSION MODELS

The three major types of erosion models based on model structure are the regression-derived, index-based, and process-based models.

Regression-Derived Models

The approach used to derive and develop an erosion model is determined by the model's structure. One way to derive an erosion model uses statistical regression procedures to fit an equation to a data set. Regression fits equations to data by minimizing the sum of the squares of the differences between the values observed and the values estimated by the equation. The equation form and independent variables (factors) in the equation are selected to give the best fit to the experimental data as measured by a statistical goodness of fit. Equations can be fitted to data, giving priority to having the percent error be the same for all soil-loss values or giving priority to fitting the large soil-loss values. The nature of the problem where the model will be applied determines whether priority is given to fitting least percent error for all soil-loss values or to fitting the large values. For example, priority is given to the relative (percent) error for each soil-loss estimate so that the model fits equally well on a percentage basis for the entire range of soil-loss estimates for models used in conservation planning. In contrast, a goodness-of-fit measure that minimizes the absolute errors rather than relative errors for high soil-loss values is used to develop models that estimate sediment yield for use in sedimentation problems where mass of sediment must be estimated accurately.

The regression-derived model is evaluated to ensure that it performs adequately over the range of conditions where it will be applied. In general, a regression-derived erosion model should not be extrapolated (applied) to conditions beyond those represented in the database used to derive the model. The adequacy of a regression-derived model is highly dependent on the quality and quantity of data from which it was derived.

Index-Based Model

Equation (5.1) is an index-based erosion model. Each variable in the equation is an index that represents the effect of that variable based on the value assigned to the index. Index-based models are also referred to as *empirically based models* because factor values are entirely derived from an empirical database. Index-based models are also referred to as *lumped process models* because the erosion processes described in Chapter 4 are not explicitly represented.

An extensive empirical database that represents a wide range of field conditions as closely as possible is required to develop an index-based model. Because an index-based model is empirical, its application is limited to the conditions represented in the database. For example, an index-based model cannot be applied to a land-use system for broccoli if the database does not include measured soil loss for broccoli land-use systems.

Soil-loss data used to derive an index-based model for interrill and rill erosion are measurements of sediment collected from plots on uniform slopes (Chapter 6). The soil-loss values are sediment yields measured at the end of the plots divided by plot length, which is the average erosion rate in mass per unit area (soil loss) from the plot. These plots measure soil loss from the eroding portions of the landscape. The index-based model is limited to the erosion processes that occurred on the plots that produced the database used to derive its factor values. The decision as to whether or not an index-based model applies to a particular field situation is based on the similarity of erosion processes at the field site to those on the plots used to derive the model.

The starting point in developing an index-based erosion model is to define a standard condition to establish a base empirical relationship between erosivity and soil loss under standard condition and measure soil erodibility. One such standard condition is a unit plot 72.6 ft (22.1 m) long on a 9% steepness that is maintained in continuous fallow (no vegetation) by periodic tillage upslope and downslope. This unit plot represents a highly erodible condition. The plot dimensions and slope steepness are arbitrary, but the length should be sufficiently long and the slope sufficiently steep for rill erosion to occur. The treatment is also arbitrary. For example, an untilled fallow soil could be used instead of the tilled fallow condition. Regardless of the standard condition, all factor values except erosivity and erodibility are set to 1 by definition—hence the name *unit plot*. A comparable standard condition for developing an index-based wind erosion model is an unsheltered (no windbreaks), isolated field with a bare, smooth, noncrusted soil surface.

The second step in developing an index-based erosion model is to establish the base relationship between soil loss from the unit-plot and erosivity. A linear relationship between erosivity and soil loss for the unit-plot condition is preferred (Figure 2.3). A linear relationship simplifies the mathematics of the erosion model so that the interaction of temporal variations in erosivity and land-use conditions can easily be represented (Chapter 2).

The fundamental part of the index-based model is therefore the soil loss computation for the unit-plot condition (A_u):

$$A_u = \text{erosivity factor} \cdot \text{soil-erodibility factor} \tag{5.12}$$

Variables used to represent erosivity are discussed in Chapters 2 and 4. The other factors in equation (5.1) for topography and land use have values that are ratios of soil loss with the given topographic and land-use condition to soil loss from the unit-plot condition. These ratios vary about 1 based on the degree that soil loss for a specific topographic and land-use condition differs from soil loss for the unit-plot condition.

The soil-erodibility factor varies during the year and by storm. An approximation that simplifies the index-based model is to use an average annual value for soil erodibility. However, the temporal interaction of erosivity and land-use condition is so important that it must be represented. Soil loss is greatest when the peak of erosivity corresponds with the land-use condition that leaves the soil most vulnerable to erosion, which is usually when cover from vegetation or applied mulch is minimal and the soil has recently been mechanically disturbed. Conversely, soil loss is least when peak erosivity occurs when cover is greatest. The relationships among soil loss, erosivity, and the factors that affect soil loss are nonlinear. Simple averages do not give sufficiently accurate estimates. The temporal distribution of erosivity and the temporal distribution of the land-use condition must be taken into account together.

Up to about 10 cover-management periods in a year are defined based on canopy development, mechanical disturbances such as grading or tillage, and changes in ground cover from harvest or placement of mulch on a construction site. Soil-loss ratio values for each cover-management period are computed from the empirical database by dividing the measured soil loss from a particular land-use condition by the soil loss from the unit plot for that time period. These soil-loss ratios vary with cover-management period as illustrated in Table 5.3. The overall land-use factor value in equation (5.1) is computed as a weighted average of the soil-loss ratios for each cover-management period based on the temporal distribution of erosivity. The equation for that computation is

Table 5.3 Cover-Management Stages, Soil-Loss Ratios, and Fraction of Erosivity in Cover-Management Period

Date	*Operation or Event*	*Soil-Loss Ratio*	*Fraction of Annual Erosivity in Period*	*Soil-Loss Ratio · Fraction Erosivity*
4/15	Moldboard-plow	0.30	0.04	0.012
5/1	Prepare seedbed	0.50	0.03	0.015
5/10	Plant corn	0.55	0.14	0.077
6/10	Cultivate crop	0.45	0.10	0.045
6/25	Approximately 50% canopy	0.28	0.40	0.031
7/10	Approximately 75% canopy	0.18	0.08	0.072
10/15	Harvest	0.056	0.08	0.005
12/31	End of calendar year	0.063	0.1	0.006
			LUF =	0.26

$$\text{LUF} = \Sigma f_i \phi_i \tag{5.13}$$

where Σ indicates a summation over the cover-management periods, f_i the fraction of erosivity in the ith cover-management period, and ϕ_i the soil-loss ratio for the ith period. Once values for the land-use factor have been computed, the values can be placed into tables, such as Table 5.4, for easy use. These computations must be made by geographic region because the erosivity distributions and dates in land-use practices vary by region.

Erosivity values and their temporal distributions are computed from long-term weather data and mapped for both water and wind erosion models. Similarly, values for the soil-erodibility factor can be assigned to soil mapping units in soil surveys (USDA, NRCS, 2001c). A value for the topographic factor is selected from a table based on the slope length and steepness or field length of the site where the erosion model is applied. Slope length is the distance from the origin of overland flow to the point that deposition begins or to a concentrated flow area if no deposition occurs. This slope length defines the eroding portion of the hillslope where a soil-loss estimate will be made. Field length is the important topographic input for an index-based wind erosion model.

The empiricism of the index-based erosion model greatly reduces its transferability to other locations, soils, and land uses unless an empirical

Table 5.4 Land-Use Factor Values for Various Locations and Cropping-Management Systems

Location	Clean-Tilled Corn	Reduced-Till Corn	No-Till Corn	Clean-Tilled Wheat
Mitchell, SD	0.22	0.11	0.008	0.097
Madison, WI	0.21	0.11	0.011	0.14
Columbia, MO	0.25	0.15	0.018	0.17
Boston, MA	0.17	0.11	0.011	0.22
Memphis, TN	0.24	0.17	0.024	0.26
Dallas, TX	0.31	0.21	0.028	0.22
Baton Rouge, LA	0.28	0.23	0.042	0.29

database is available to compute the factor values for those locations. If the database does not exist, the development process described above, including the experimental measurements, must be repeated. The difficulty of extending the index-based erosion model to situations not represented in the database used to derive its factor values is a major limitation. The strength of the index-based model is its simplicity, ease of use, and power relative to its complexity.

Simple Process-Based Models

The simple process-based models are similar to the index-based models but with one major difference. The simple process-based erosion model for water erosion is divided into two components, one to represent interrill erosion and one to represent rill erosion. The major reason for developing the simple interrill–rill process-based model is that interrill erosion behaves differently with respect to the controlling variables than does rill erosion (Chapter 4). For example, ground cover has less effect on interrill erosion than on rill erosion, as illustrated in Figure 4.6 (Foster, 1982). Also, slope length has no effect on interrill erosion, whereas rill erosion increases with slope-length. The index-based model assumes a single slope-length relationship for all conditions, which does not take into account how the slope length effect varies as the ratio of rill to interrill erosion varies for various climatic, soil, topographic, and land-use conditions.

A simple process-based model for water erosion is the same as equation (5.1) except that it has separate terms for interrill and rill erosion. The structure for the two terms is much like the structure in equation (5.1); that is,

$$\begin{aligned} \text{SL} &= \text{interrill soil loss} + \text{rill soil loss} && (5.14)\\ &= (\text{IE} \cdot \text{ISEF} \cdot \text{ITF} \cdot \text{ILUF}) && (5.15)\\ &\quad + (\text{RE} \cdot \text{RSEF} \cdot \text{RTF} \cdot \text{RLUF}) \end{aligned}$$

where IE is the interrill erosivity, ISEF the interrill soil-erodibility factor, ITF the interrill topographic factor, ILUF the interrill land-use factor, RE the rill erosivity, RSEF the rill soil-erodibility factor, RTF the rill topographic factor, and RLUF the rill land-use factor. In addition to the process-based soil-loss equation shown as equation (5.15), a simple process-based equation is developed to estimate soil loss when deposition occurs on the rill areas. This deposition occurs in furrows, for example, when ridge–furrow systems are on a flat grade, less than about 2%, because the transport capacity of the flow in the furrow is less than the sediment load delivered from the ridge sideslopes (Chapter 4). That deposition model has the form

$$D_p = (V_f/q) \cdot (T_c - G)/\lambda \qquad (5.16)$$

where D_p is the deposition, V_f the fall velocity of the sediment in still water, q the flow rate, T_c the transport capacity of the flow, G the sediment load in the flow, and λ the slope length (Renard and Foster, 1983). The equation for soil loss is

$$\text{SL} = \text{interrill soil loss} - D_p \qquad (5.17)$$

Equation (5.16) computes little deposition when the sediment particles are fine, represented by a low fall velocity, and much deposition when the difference between transport capacity and sediment load is large. This difference increases as the transport capacity decreases (e.g., from a decrease in land slope) or as sediment load becomes large because of high interrill soil loss from ridge sideslopes. Equation (5.16) and (5.17) require equations for computing transport capacity as a function of flow rate, slope steepness, and hydraulic roughness (Chapter 4).

A comparable simple process-based erosion model for wind erosion is

$$\text{SL} = D_s + D_a \qquad (5.18)$$

where D_s is the soil loss by shear stress that the wind exerts on the soil and D_a is soil abrasion by windblown sediment striking the soil surface (Chapter 4). A model for deposition of wind-transported sediment is essentially the same as equation (5.16)—except for fall velocity being for still air.

Combined Index- and Process-Based Models

The index-based model requires data for specific land-use systems to develop factor values so that the model can be applied to those systems. For example, if measured soil-loss data are not available for no-till broccoli, the model cannot be used to develop a conservation plan for a farmer wishing to grow broccoli with a no-till land-use system. A new type of erosion prediction technology was developed that combined the best features of the index-based model and the simple process-based model to meet the greatly expanded use of erosion prediction that began in the early 1970s and accelerated with implementation of the 1985 Food Security Act (Chapter 8).

The starting point for developing this type of erosion model is the structure of equation (5.1). A major substitution and addition is a set of equations to compute temporal values for the land-use factor. These equations use variables that represent the basic features of land-use systems that affect erosion. These variables, known as *cover-management subfactors*, include canopy cover, height, open space, and height from which water drops fall from the canopy; standing stubble; ground cover; surface roughness; buried plant residue; live and dead roots; time since last mechanical soil disturbance; ridge height; and orientation of ridges to the prevailing direction of wind or surface runoff. These subfactor equations compute the effect of land use on soil loss for any land use, which include conditions well beyond those represented in the empirical database used to develop the index-based model.

The index-based model uses a single relationship to describe the effect of slope length on soil loss, but theory and experimental data show that the relationship varies with the ratio of rill to interrill erosion (McCool et al., 1989). The empirical structure of the index-based model is retained, but a simple process-based model is added to adjust the empirical slope length relationship as a function of the ratio of rill to interrill erosion.

Experimental data also show that the effectiveness of ground cover also varies. Both theory and experimental data show that the variation can be explained by the ratio of rill/interrill erosion in the absence of ground cover. A rill/interrill erosion ratio computed with a simple process-based model can be used to adjust the *b* value in the empirical relationship, equation (5.6), for the effect of ground cover in the index-based erosion model.

The index-based model cannot be applied to modern conservation tillage used in conjunction with buffer and filter strips (narrow strips of permanent grass) and stripcropping (Chapter 7) because experimental data are not available to derive factor values for these practices. A deposition equation such as equation (5.16) is added in the index-based model to compute a sediment delivery ratio, which in turn is used to compute soil loss for modern stripcropping and narrow grass strips. This same simple process-based deposition equation is used in the index-based model to compute soil loss, deposition, and sediment yield values for hillslopes having concave sections where deposition occurs.

The time step used in the index-based model to compute the interaction of the temporal variations in erosivity and the land-use factor is based on cover-management periods that vary in duration from a few days to several weeks. In the combined index-process based model, a daily time step is used to compute average annual daily soil-loss ratio values for land-use effects. Also, equations are developed to compute the average annual daily variations in the soil erodibility and topographic factors. The proper mathematical solution to the equations is to compute a soil loss on each day and sum them to obtain an annual value instead of the approximate mathematical method used in the index-based model.

The fundamental equation used in the index-based model is

$$\mathrm{SL} = \mathrm{CF} \cdot \sum f_i \kappa_i \cdot \mathrm{TF} \cdot \sum f_i \phi_i \tag{5.19}$$

where κ_i is the soil-erodibility factor for the *i*th cover-management period, $\mathrm{SF} = \sum f_i \kappa_i$, and $\mathrm{LUF} = \sum f_i \phi_i$ in Equation (5.1), CF is the climatic factor, and TF is the topographic factor as defined for equation (5.1).

The fundamental equation used in the combined index-process based model is

$$\mathrm{SL} = \mathrm{CF} \cdot \sum f_i \cdot (\kappa_i \cdot tf_i \cdot \phi_i) \tag{5.20}$$

where tf_i is the the topographic factor for the *i*th period. Equation (5.19) is where the variables are first averaged and then multiplied. Equation

(5.20) is where the values are first multiplied and then summed. The difference in soil estimates between equations (5.19) and (5.20) can approach 20%. Equation (5.19) is an approximation used in early index-based models for simplicity so that the model could be used easily in conservation planning in the field office. Equation (5.20) is the proper mathematical treatment of the variables, but this formulation is too difficult to solve without using a computer. Simplicity was necessary when erosion models first began to be used in the 1940s before computers became widely available.

The simple inputs of the index-based model can be retained in the combined index- and process-based model. This model combines the best of index-based and simple process-based models, but the combined model must be run on a computer. The simple approach of substituting values for each factor in equation (5.1) and computing soil loss with a simple equation is lost with the combined model. However, the power and the conditions where the combined index and process-based model can be used significantly exceed that of the index-based models.

Dynamic Models

Erosion occurs as a series of discrete events with different erosion amounts for each event because of differences in storms and land-use conditions at each event. Erosion varies during a storm because rainfall intensity, infiltration rate, runoff rate, and wind velocity vary greatly during storms. Erosion is dynamic on two temporal scales. One scale is the days and weeks between storms associated with plant canopy development and loss of residue cover by decomposition. Events such as mechanical soil disturbance, harvest, and placement of mulch cause abrupt, step changes in roughness and cover on the dates of these events. The other scale is the minutes and hours during an erosion event.

Dynamic erosion models track temporal variables by computing values at regular points through time between storms events. These models use computed values at the time of the storm for variables such as canopy, ground cover, and roughness. These dynamic models are also referred to as *simulation models* because they mathematically "simulate" field processes through time. An advantage of dynamic erosion models is that they compute year-to-year variations in erosion. For example, these models compute soil loss for dry years that take into account the possibility of a very erosive period following a dry period when plant canopy and ground cover are reduced. Soil loss is greater than would be computed using average annual values like those used in the index-based models. The reverse is true when erosivity is low following a wet period. Erosion models that use average annual daily values overestimate soil loss for this condition. To a certain extent, the over- and under estimates average out, but not entirely. The major reason for computing soil loss for each storm is that the mathematical equations are nonlinear. Computing soil loss by individual storms produces a more mathematically correct estimate than using average annual daily values because of these nonlinearities.

Most dynamic models use similar procedures to compute values for the periods between storms. However, dynamic models differ greatly in the way they compute soil loss for the erosion event. One type of dynamic erosion model uses event duration and representative values for rainfall intensity, runoff rate, or wind velocity for the storm in steady-state equations to compute soil loss and other erosion variables for each storm. These models are referred to as *quasi-steady-state models* because they use the steady-state conservation-of-mass equation rather than solving the dynamic conservation-of-mass equation throughout the storm.

The other type of dynamic erosion model uses the dynamic conservation of mass equation. This model computes erosion variables at regular intervals during the storm using values for rainfall intensity, runoff rate, or wind velocity as they vary during the storm. The benefit of making these detailed computations within a storm results from nonlinearities in the governing equations and rapidly changing conditions during the event. The quasi-steady state models do not capture the effects of changes within a storm. For example, transport capacity and deposition may be limiting during the early part of an erosion event, detachment limiting in the middle of the event, and transport capacity and deposition limiting during the latter part of the event. Quasi-steady state models assume a single case of detachment or transport limiting for the entire storm and therefore are less powerful than the fully dynamic models.

The effect of nonlinearities on soil loss computations can be illustrated by computing detachment by raindrop impact on interrill areas. This detachment rate is proportional to the square of rainfall intensity (Foster, 1982; Meyer, 1981). The equation that computes interrill detachment is:

$$D_i = k_i i^2 \tag{5.21}$$

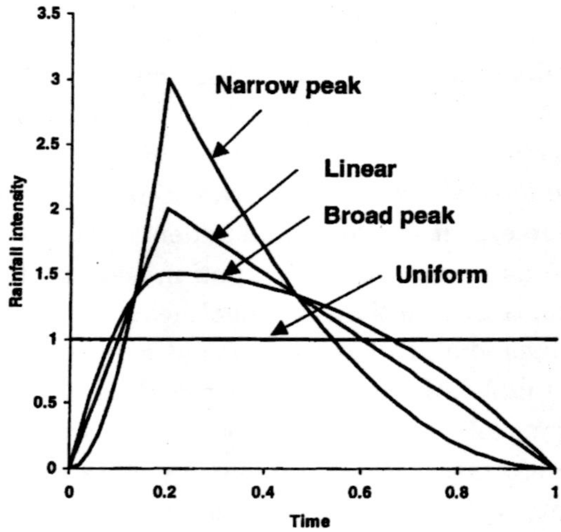

Figure 5.2 Rainfall intensity patterns.

where D_i is the interrill erosion rate, k_i the interrill soil-erodibility factor, and i the rainfall intensity. Figure 5.2 show four rainfall-intensity patterns where the average rainfall intensity is the same for each pattern. The effect of nonlinearity in equation (5.21) is evident in Table 5.5. Total storm detachment is different for each rainfall intensity pattern even though average rainfall intensity is the same for all the patterns. Therefore, the representative rainfall intensity used in the quasi-steady-state model depends on the rainfall intensity pattern. As the rainfall intensity pattern becomes more narrowly peaked, the representative rainfall intensity used in equation (5.21) to compute an average erosion rate for the storm increases. A maximum rainfall intensity averaged over 30 minutes frequently is used in erosion models based on empirical analysis of soil-loss data (Chapter 2, Wischmeier, 1959). The results of Table 5.5 shows that the duration over which to compute a maximum intensity varies as a function of the rainfall pattern, which varies by location and season (Brown and Foster, 1987; Richardson et al., 1983).

The fully dynamic models have the greatest power of all erosion models. For example, they account directly for the effects of nonlinearity, as illustrated above for interrill erosion. They also take into account changes between detachment and transport-limiting conditions throughout a storm. The disadvantage of these models is their great complexity and their reduced robustness. *Robustness* refers to the degree that the model can be extrapolated without computing erroneous values that to do not

Table 5.5 Effect of Rainfall Pattern on Detachment on Interrill Areas[a]

Rainfall Pattern	*Total Detachment for Storm*	*Representative (Effective) Intensity*	*Peak Intensity*	*Fraction of Storm Duration to Average for Representative Intensity*
Uniform	1.00	1.00	1.0	1.0
Broad	1.20	1.10	1.5	0.9
Linear	1.34	1.16	2.0	0.8
Narrow	1.81	1.35	3.0	0.7

[a]Average intensity = 1.

follow expected trends and the likelihood that a combination of ordinary events in the model cause significant errors. Also, the computer programs for these models require longer "run time" than do computer programs for simpler models, which is a major drawback in conservation planning at local field offices. The fully dynamic models most often are used in research applications. The potential performance of the dynamic models that use representative storm values falls between the performance of fully dynamic models and the performance of combined index and process-based models. A major limitation of both types of dynamic models is the large number of variables. Measurements of certain variables can be difficult and costly, and input values may not be readily available. Less than the full potential of dynamic models typically is realized in field applications.

OTHER TYPES OF EROSION MODELS

Stochastically Driven Models

All erosion models require input values for weather variables, such as rainfall, wind, and temperature, to "drive" the equations that compute runoff, soil moisture, erosion, and other variables. The index-based and and combined index- and process-based erosion models use average annual monthly and biweekly weather data inputs. The combined index-based model divides the monthly values into average annual daily values for use in the erosion equations. Although the dynamic models can use historical weather data for input, a *weather generator* is typically used to compute weather input values. A weather generator is a set of mathematical equations that compute weather values having the same statistical properties as the historical data (Johnson et al., 1996). Measured weather data at particular locations are analyzed to determine input values for the weather generator.

A weather generator operates like drawing numbers randomly from a jar. The measured weather data are analyzed, and the rainfall amounts for each rainstorm are recorded on a piece of paper. The next step is to determine whether rainfall is likely to occur on a particular day. The likelihood of rainfall occurring on a given day depends on whether rainfall occurred on the preceding day. Thus, two jars are used for the likelihood of a rainfall event on a given day. One jar is for the case where rainfall occurred on the preceding day (a wet day), and the second jar is for the case where rainfall did not occur on the preceding day (a dry day). Each jar would be filled with pieces of paper with a zero (0) or one (1) written on them. The proportion of zeros to ones represents the probability of a rainfall event. If a zero is pulled from a jar, no rainfall is predicted to occur, and if a one is pulled from a jar, rainfall is predicted to occur on that day.

If rainfall is predicted to occur, a piece of paper is pulled from the third jar, containing the rainfall amounts. Most of the rainfall amounts in the jar are small. Only a very few of the rainfall amounts are large, which in statistical terms means that the probability distribution curve for rainfall amounts is highly skewed toward small rain amounts (Richardson, 1981). That is, far more small rain events occur than large ones. Large rainfall and wind events produce a disproportionate amount of soil loss (Edwards and Owens, 1991; Hjelmfelt et al., 1986; Larson et al., 1997). Therefore, the accuracy of the average annual soil-loss estimate depends greatly on the accurate generation of a few large events. A database covering as many years of rainfall and wind as possible is needed to develop the input values for the weather generator, but frequently the database is no more than a few years in duration. A short weather record leads to poorly defined input values and large errors in estimates of soil loss caused by large, infrequent rainfall and wind events. Another limitation of some weather generators is the lack of cross correlation between variables. For example, a weather generator should not compute high solar radiation on days on which rainfall occurs.

Although these limitations do not invalidate the use of weather generators, they degrade the performance of erosion models so that the performance of a detailed, complex model may not be much better than the performance of simple models when used for conservation planning at local field offices.

Spatially Distributed Models

Detachment, deposition, and sediment load vary greatly over space, even along a uniform slope where the net detachment rate at the end of the slope length is about 1.5 times the average for the entire slope length. As Figure 2.4*b* shows, the variation in net detachment and net deposition by water erosion on a complex slope can be even greater than on a uniform slope. Soil loss and associated variables also vary over a region, as weather, soil, topography, and land use vary. Because the mathematical equations are nonlinear in space (e.g., distance along the flow path), the accuracy of soil loss, deposition, and sediment–yield estimates is improved by solving the equations over space using spatially distributed input values rather than spatially averaged values.

Most erosion models, regardless of type or structure, can be applied in a spatially distributed way. Simple erosion models can be applied to compute soil loss at sample points over a region (Chapter 6). These computations provide a spatially distributed set of soil-loss values by location within the region. These values can be aggregated to compute average soil loss by soil type, land use, or other important variables.

If the interest is sediment yield from regions where deposition occurs, the flow and sediment must be routed by solving the conservation-of-mass equation over space. A key element is the topographic description of the landscape, which usually is a contour map or a set of spatially distributed digital elevation values, In general, flow paths for water erosion are defined from the topographic data, and the erosion equations are solved in a stepwise fashion along the flow paths starting at the upper end of each overland flow path. A similar approach is used to compute wind erosion considering that the flow paths change with wind direction. The starting point for the wind-erosion computations is the location where the sediment load is either zero or a known value.

Spatially distributed models are frequently structured for geographical information systems (GIS) and other technologies designed to handle large amounts of spatially distributed information. However, even with these technologies, a large number of input values must be developed for a spatially distributed model applied to a large area. Many of these data are increasingly available in computerized databases such as the USDA, NRCS soil-survey database (USDA, NRCS, 2001c). Models can read directly from these databases to save much time and effort in acquiring and entering data. Still, using spatially distributed models over large regions is time consuming, including computer run time, especially if the model is fully dynamic.

Spatially distributed models have great power. However, obtaining improved results from spatially distributed models depends on both the vertical and horizontal resolution of the input topographic information. If resolution of the topographic data is not sufficient to define slope shapes such as concave segments, the model cannot possibly compute the deposition that occurs on concave slope segments regardless of the scientific sophistication of the model. A simple model with well-chosen inputs almost always out performs a highly detailed model with poor topographic and land-use inputs.

STEPS IN DEVELOPING AN EROSION MODEL

The best erosion models are those developed cooperatively with users working alongside model developers. Developing a simple research model is a relatively small project in comparison to developing an erosion model that is to be used by a national agency in conservation planning for many locations, soils, topographies, and land uses. Such large projects require effective project management to be successful. A series of systematic steps are followed in a project to develop an erosion model: (1) develop a set of user requirements, (2) select the underlying structure of the model, (3) design the computer interface, (4) develop the governing equations, the linkage among them, and parameter values, (5) develop a core database, (6) verify, (7) validate, (8) evaluate, (9) expand databases, (10) document, (11) implement, and (12) upgrade over the life cycle of the model.

Developing User Requirements

Except for the model itself, the user requirements are the most important part of the model development project. This document states the users' expectations for the model and how the model will be used. The most important statement is the purpose of the model because this statement will be used to determine the validity of the model. The user requirements also specify the desired inputs and outputs for the model, the environment in which the model will be used, resources available to use the model, and required accuracy for the model outputs. The user requirements may also describe other models, computer programs, and databases with which the model must interface and share components. The user requirements may specify that the model use a comparable structure as another model and produce the same or similar values for certain variables as produced by another model. The user requirements may specify the type of model. Users typically expect that the model will be easy to use, require minimal resources, require only readily available input values, be easily understood, and produce highly accurate results for an extensive set of output variables.

Selecting the Model Type

The single most important decision with respect to the model itself is selecting the model's type and associated structure. Model structure largely determines the types of equations used in the model, the experimental data required to develop the model, the type and extent of input data, available outputs from the model, and the resources that will be required to implement and apply the model. Each model structure has advantages. A model structure that works well in a research application may not work well for conservation planning at the local field office. Robustness is less important in research applications but is very important in conservation-planning applications.

Designing the Computer Interface

Almost all modern erosion models are implemented in a computer program. The computer interface, which includes the screen appearance and computer mechanics, is the connection between the user and the model. The user enters inputs and receives outputs from the model through the interface. The interface should be easy to use, employ operations that are obvious and consistent with common practice, and display outputs that are clear and immediately understandable.

The interface should be designed based on user requirements using a top-down approach before the model itself is designed. A typical interface has layers. In a top-down design, the top layer, where the inputs are entered for location (climate), soil, topography, and land use, is designed first. Users "drill down" to lower interface layers to enter data for details of climate, soil, topography, and land use. Models are usually programmed in modules to simplify model maintenance and updating.

Developing the Governing Equations, Their Linkage, and Parameter Values

The governing equations, their parameter values, and the logic and computational sequence used to execute the equations are the heart of any erosion model. However, the type of model determines how the equations are derived, the experimental data required to develop the model, and how the data are analyzed to determine parameter values. The regression-derived model is developed using regression analysis to fit an equation directly to data. The process of fitting the equation to the data determines the parameter values for the model. A database is required with values for the dependent erosion variables of interest, such as soil loss, deposition, sediment yield, and sediment characteristics, along with values for the corresponding independent variables that affect erosion, such as soil texture, canopy height, surface roughness, and ground cover.

The index-based model requires a database of soil-loss measurements for a wide range of land-use practices, such as the cropping and management systems used on cropland. Values for each factor in the index-based model are derived by analysis of this database. The experiments used to produce the data contain a reference (standard) treatment, such as the tilled fallow treatment, and treatments representing the land-use systems to which the model will be applied.

In the combined index- and process-based model, two sets of data are collected. One set is the same as that collected to develop the index-based model. The second set of data is used to derive the equations that describe how subfactors such as canopy, ground cover, roughness, ridges, degree of soil consolidation, below ground biomass, and their interactions affect soil loss. The form of these subfactor equations is either derived empirically or based on erosion theory. The first data set is used to evaluate how well the subfactor equations describe soil loss for particular land-use systems. Data from the land-use systems can also be used to refine the subfactor equations and their parameter values.

The governing equations for a process-based erosion model are derived from erosion theory and experiments for individual erosion processes. The types of experiments and measurements are different from those used to develop subfactor equations in the combined index- and process-based model, where measurements were made without explicit regard to erosion processes. A data set for land-use systems such as that for the index-based model is used to ensure that the process-based model performs adequately when applied to field situations. The database for land-use systems is not used to develop equations and parameter values but to validate and refine the equations and parameter values developed from experiments for individual erosion processes.

Some process-based erosion models are calibrated to the data set, where soil loss was measured directly for land-use systems. Calibration involves systematic adjustment in parameter values so that the model gives the best fit to the experimental data. Great care must be exercised in calibrating an erosion model. This approach frequently leads to questionable parameter values and reduced robustness (Foster, 1980). Most erosion models have a large number of variables, and calibration is analogous to using regression to fit an equation to data where the number of independent variables is almost as large as the number of data points. Erosion data often contain a high degree of unexplained random variability (Nearing et al., 1999). Sometimes the model is calibrated to noise in the data rather than to the main effects of the independent variables. Also, erosion processes do not behave as distinctly and without interaction in the field as assumed in the model. Not all effects are represented in any model, which requires distortion in the model's parameter values to represent the missing effects. That is, the parameter values required for the model to fit the combined data from interrill and rill erosion are not those that fit the individual processes when measured separately. Developing robust equations and parameter values that reliably capture the main effects of the most important variables across a wide range of conditions is part of the art of developing process-based erosion models.

Developing the Core Database

Users enter values in the erosion model to represent a specific site and the land-use systems involved in the particular application of the model. These input values must be based on the same definitions and measurement techniques as those used to develop the model. A procedure that ensures this consistency involves developing a core database from the research input values used to develop the model. The core database covers major types of vegetation and operations (events that change vegetation, plant residue, or the soil). The core database is expanded into a full operational database. Values added to the operational database must be consistent with the values for similar conditions in the core database. Suppose that input values are needed for a land-use system involving watermelons but that the required research data on watermelons are not available. Because watermelons are much like cucumbers, values for cucumbers in the core database are adjusted to reflect differences between watermelons and cucumbers.

Verifying the Model

Verification is the process of ensuring that the model makes the calculations as intended. Verification ensures that the equations, parameter values, and logic that links the equations have been programmed as designed and give expected results. Verification involves running the model for the range of research data used to derive the model, the core database, and field conditions for which the model might be used. Also, verification involves running the model for special conditions to make sure that every equation and every logic step in the model is exercised. The objective is to test every element of the model to find and fix all errors.

Validating the Model

Validation is the process of ensuring that the model serves its intended purpose as described in the user requirements. Although an important part of validation is to determine how well the model fits measured data, this step is only one of several evaluations that are made and is only one of several considerations used to determine if the model is valid.

Erosion models typically fit measured average annual soil loss with an uncertainty of about ±25% for moderate erosion rates of about 3 to 30 tons/acre (6 to 60 metric tons/ha) per year and increases to ±50% for values greater than about 50 tons/acre (10 metric tons/ha) per year and to errors as large as 1000% for small loss values of less than 0.1 ton/acre (0.2 metric tons/ha) per year (Risse et al., 1993; Wischmeier and Smith, 1978). The objective is to fit measured data as closely as possible, provided that the data appropriately represent field conditions. Most data used to determine model performance are collected under research conditions, but research conditions may not represent field conditions. An example is the effect of fabric fences used to control sediment delivery from construction sites. Fabric fences are highly effective at trapping sediment in the laboratory, but they often do not perform well in the field because of poor installation and maintenance. An erosion model that fits laboratory data well for fabric fences may not provide good estimates when applied to actual field conditions.

Another situation that arises during model development and validation is whether to give priority to fitting all values with the same percent error or to give priority to fitting the large values with minimal error. Erosion models used in conservation planning should be most accurate in the range 0.5 ton/acre (1 metric ton/ha) per year to 20 tons/acre (40 metric ton/ha) per year. Any estimated soil loss greater than 20 tons/acre (40 metric ton/ha) per year is considered excessive regardless of whether the true value is 20 or 100 tons/acre (40 or 200 metric tons/ha) per year. Similarly, a soil-loss values less than 0.5 ton/acre (1 metric ton/ha) per year is considered low regardless of whether the value is 0.5 ton/acre (1 metric ton/ha) or 0.01 ton/acre (0.02 metric ton/ha) per year. The critical question is whether or not the model is sufficiently accurate to justify a land user reducing soil loss from 8 tons/acre (18 metric tons/ha) to meet a soil-loss tolerance of 5 tons/acre (11 metric tons/ha) per year when a significant investment is required. Erosion models used to design reservoirs, for example, should most accurately estimate large values, because these values determine the reservoir design. Values for low sediment concentration of fines may be critical for a model being used in air- and water-quality analyses.

Erosion data are highly varied, which is the reason that replicates are used to measure soil loss. For example, the difference in soil loss between replicated plots located less than 20 ft (5 m) apart and prepared identically can be 30% for a measured erosion rate of 10 tons/acre (20 metric tons/ha). The percentage difference is far greater when soil loss is very low. For example, the soil loss from one replicate can be 0.1 ton/acre (0.2 metric ton/ha) and 0.5 ton/acre (1.0 metric ton/ha) on the replicate plot for a 400% difference. The important issue is not the difference between replicates but the differences between treatments. If the differences between replicates are too large, too few replicates are used in the experiment, and the difference between treatments is too small, the hypothesis cannot be rejected that the treatments are not different. However, regardless of the differences between replications in this example for these two treatments, one treatment is clearly superior to the other treatment for controlling soil loss. Soil loss must be interpreted in this context when developing a model for conservation planning. A model cannot be expected to fit the data more closely than the random variability in the data. A model's poor fit to experimental data may indicate errors in the data rather than problems with the model. The purpose of the model is not to explain differences between replicates but differences between treatments.

An erosion model that fits the research database well does not ensure that the model is adequate for all field conditions. All research erosion databases are limited, even those that contain thousands of plot years of data. Professional judgment is used to determine if the model is satisfactory for all the ways that it will be used. An evaluation technique is to determine whether the model is correctly computing the main effects and trends associated with key variables, such as ground cover, based on scientific knowledge and the judgment of erosion scientists and technical specialists.

Even if a model makes all its computations in the best way possible, the model still may not be valid. Use of any model requires resources, including skill of the user, time, input data, computer hardware and software, and technical support. The performance of any model improves as the expertise of the user improves. Access to user guides, technical documentation, training, and other users are key factors in applying a model successfully. The model should be easy to use, which is often a function of the model's computer interface—but not entirely. Also, the time required to use a model should be minimal. Ultimately, the time and difficulty in using a model is weighed against the benefits of using the model. The consideration is not in the context of how well the model fits data but how well the model serves its intended purpose, such as a guide to conservation planning. Does the conservation plan improve sufficiently when an alternative model is used in proportion to the increase in resources required to use the alternative model? If two models result in the same conservation plan, each model performs equally well, and choice of the model is based on preference and resources required to use the model.

A model may give great results, but if it requires a powerful computer, extensive computer run time, or data that are not available or cannot be obtained without significant expense, the model will not be used. Therefore, the model is not valid because it cannot serve its intended purpose. Conversely, a certain model is sometimes used in nonrecommended ways because the user judges that the results are satisfactory for his or her purpose, or the nonrecommended model has certain advantages, such as using easy-to-obtain input data. An example is used if a model intended to estimate average annual soil loss to estimate soil loss from a single storm.

Evaluating the Model

Before a user implements a model for production work, the user should evaluate the model thoroughly. The user environment is very different from the research and development environment. A model that works well in a development environment may not work well in the field environment. Users evaluate the model at this stage and provide a formal statement to the model developers concerning changes and revisions that are required. This step is the last opportunity for the user to have a major impact on the model. The objective is for the model to work well at this stage rather than to make major changes. Major changes become increasingly difficult and costly as the model development progresses.

Expanding the Databases

An extensive set of databases is needed to apply an erosion model in field applications such as conservation planning. These databases include information on the weather and soils for the region where the model will be used. Data on vegetation, plant residue, and operations are universal for all locations in some erosion models, but some data may require customizing for specific regions. Input data on land-use systems must be developed. This information includes dates of operations, names of operations, and the vegetation or residue associated with particular operations such as planting and adding mulch. When the user runs the model, a land-use system name is entered, and the data stored in the database for that land-use system name are used in the model to compute soil loss.

The field-application database is developed by expanding the core database using a specific set of rules, definitions, measurement techniques, and other procedures involved in assigning values to input variables based on the requirements of the particular model. Similar variables are used in various hydrologic and erosion models, but definitions of variables should be studied carefully before adopting values from a database for another model.

The importance of consistency in the database used for field applications of the model across locations, operations, vegetations, residues, soils, and land-use systems cannot be overemphasized. Also, measured data need to be inspected carefully for missing values, errors, and differences in length of record for weather data.

Documenting the Model

Documentation is an important part of model development and use. The most important part of the documentation is a user guide that describes how the model works and how the model should be used. The user guide describes the key input variables, their definitions, and procedures for choosing values for them. Another important part of the user guide is information on how to set up the model to analyze particular types of problems and how to interpret outputs. The computer program should contain extensive "help" information. The interface of the model should be designed so that the model can easily be used, and steps to operate the model should be obvious without the user having to read detailed instructions.

Another important part of the documentation is a technical description of the equations, parameter values, and logic used in the model and how the model fits measured data. This information should be sufficiently complete that scientists and technical specialists can review the model and judge the scientific and technical adequacy of the model. This information should also be sufficiently complete that other model developers could reproduce the model.

The other important part of the documentation is information concerning the computer program, including the interface. The computer program should be written according to industry standards with much internal documentation. High-quality documentation greatly facilitates maintenance of the model's computer program.

Implementing the Model

Users of erosion models range from individuals to large consulting firms and government agencies. Large organizations must put in place numerous support services before the model is implemented. The organization must review the computer program that runs the erosion model to determine that the program meets organization standards. The required computer hardware, software, and databases must be in place on local computers where the model will actually be used. Documentation, especially user guides, must also be in place. Using the model must be learned, which may require formal training. The model must be integrated into the organization's operating procedures to ensure that the model is being used as desired to accomplish the organization's objectives. A support system is needed to assist users. A system is needed to work with model-maintenance personnel and model developers to correct errors and upgrade the model. The model, databases, and documentation must be distributed, often over the Internet (Appendix C).

Upgrading over the Life Cycle of the Model

Erosion models used routinely by large organizations tend to have a long life. Revisions are made periodically without fundamentally changing the structure of the model, the input database, and the values produced by the model. Making major changes requires a major investment of resources, including training, updating documentation, and dealing with computed values that are now different from values computed previously, which can have major policy implications in a conservation agency. Thus, organizations are slow to make changes in models. Are the improvements gained by changing a model worth the investment required to change? Organizations evaluate the benefits of changing a model in the context of their program objectives. If the conservation plans developed with an old model are essentially the same as those developed with a new model, change is not likely.

However, minor changes in a model are common over the life cycle of a model. Planning for, managing, and implementing minor changes throughout the life cycle of the model should be an integral part of the model development process. Model maintenance must be an important consideration during model development to minimize later costs incurred with model maintenance. Although, changes are typical, high-quality initial design is strongly preferred so that "it is done right the first time," rather than fixing problems later. This requirement does not preclude developing a model where features will be added later, but the model should be designed so that features can easily be added.

CHOOSING A MODEL

Organizations and individuals often find themselves choosing an existing erosion model without having been involved in the development of the model. The first step in choosing a model is to identify how the model will be used, the information that is needed from the model, and the information that is available for input into the model. The second step is to determine the resources available for implementing and using the model in comparison to the resources required to implement alternative models. Resources must be available for installing the model, training, collecting input data, running the model, and interpreting model outputs. A more powerful model might be preferred, but the resources may not be available to implement and apply it. The third step is to review the available models and select one. Selecting a model is always a compromise, partly because no model does everything exactly as desired. The selection process involves listing priorities and considering ways of overcoming the shortcomings of each candidate model. Factors such as ease of use, readily available input data, and acceptance by the scientific and technical community are major considerations. Although accuracy is always a consideration, accuracy must be considered in the context of the application and whether or not the expected outcome is achieved. Does the model lead to the desired conservation plan? What is the required accuracy, and is the increased accuracy worth the increased costs?

An important consideration in selecting a model is whether an established user base exists for the model. If the user base is large for a particular model, the model is probably well accepted and issues related to accuracy resolved. Also, large, high-quality input databases are probably available that can be obtained to greatly reduce the effort of developing input databases. If the model has been used for some time, the rough edges of the model have probably been smoothed and many of the problems typical of a new model fixed. Also, technical support is more readily available when a widely used model is selected. However, a consideration in choosing a model supported by another organization is whether changes and modifications can be made to customize the model to your application. Can the model be modified? If the model can be modified, who will make the modifications, and at what cost?

After a model is selected, the model must be implemented. The considerations on implementation discussed above as part of the development process also apply to implementing an existing model.

SENSITIVITY ANALYSIS

Sensitivity of a model refers to the degree that soil loss, deposition, sediment load, or other important output variables change per unit change in the input values for key variables such as slope length, field length, slope steepness, soil erodibility, ground cover, root biomass, roughness, ridge height, and orientation of the ridges. A sensitivity analysis involves inputting a range of values for a particular variable and observing the changes in values for computed variables. The input variables that cause a great change in the output for small changes in input values are sensitive variables. Increased attention should be given to selecting input values for the sensitive variables because of their great effect on the output.

A sensitivity analysis is a good way to learn a model and how it responds to inputs. The sensitivity analysis should include the conditions where the model will be used and special conditions to test the fundamental behavior of the model. Reviewing output from the model shows how the model behaves and whether the results are as expected. An error may even be discovered. Errors are not expected, but modern erosion models are complicated, and when a model is applied to a new situation, errors are sometimes found. Although the model developers should verify and validate the model, users should conduct sensitivity analysis to ensure for themselves that the model gives acceptable results.

Results from sensitivity analyses must be interpreted carefully to avoid erroneous conclusions because sensitivity often depends on the situation. In one case, soil loss may not be sensitive to a particular variable but be highly sensitive in another case. Soil loss is not sensitive to slope length at slope steepness below 1% because soil loss is hardly affected by slope length at low steepness. However, soil loss is moderately sensitive to slope length on hillslopes steeper than 20%. Ground cover is one of the single most important variables affecting soil loss, but the sensitivity of the model to ground cover depends on the situation. Soil loss is most sensitive to ground cover when mulch is applied to smooth, bare construction sites. Crop residue cover at planting is used as an indicator for the level of erosion reduction provided by conservation tillage in cropping systems. Sometimes the sensitivity to residue cover is not as great as expected for conservation tillage when compared with applying mulch to a construction site. The mulch is essentially the only land-use subfactor affecting soil loss on the construction site, but many other subfactors, including roughness, buried residue, and live and dead roots, affect soil loss with conservation tillage. These other variables decrease the sensitivity of the estimated soil loss to changes in ground cover. Buried residue can reduce soil loss significantly based on the concentration of biomass in the upper soil layer. This concentration is determined partly by the depth of tillage.

The greatest effect of this biomass is when it is buried in a shallow surface layer. Estimated soil loss is not sensitive to the depth of incorporation by a secondary tillage operation that follows a primary tillage operation such as a moldboard plow, which deeply buries most of the residue. Changing the depth of incorporation of the secondary tillage operation has little effect on soil loss when it follows an operation that buries most of the residue. However, if the secondary tillage operation occurs alone, depth of incorporation has a very significant effect on soil loss, assuming a significant amount of biomass is buried.

SUMMARY

More than a half century of use has proven that mathematical erosion models (erosion prediction technology) are very powerful tools for guiding conservation planning, inventorying soil erosion, estimating sediment yield, and use in sedimentation and air- and water-quality analyses. Var-

An erosion model should be based on adequate science, but a model based on sophisticated science does not always ensure that the model works well in field applications. Also, poor-quality inputs ensure poor results from any model. The best science-based model cannot overcome poor-quality inputs. Often, models based on high-level science are less forgiving of poor input than are less sophisticated models. Both model development and application is an art that grows with understanding of the model and with the experience of applying the model. Skill, art, and experience can overcome limited but adequate science, but poor art defeats the best science.

The criterion used to judge models is based on how well the model serves its intended purpose. Some typical requirements for validity include that the model be easy to use within available resources, that the model be based on accepted science, and that accuracy of the estimates computed by the model is appropriate for the intended application. Models should be used carefully. They should always be used as a guide to decision making; models should not dictate the decision. People, not erosion models, should be in charge.

See Appendix C for sources of information on specific models.

ious types of mathematical models are available, each with particular strengths and limitations. An erosion model should meet several requirements. The model should provide the desired information, be easy to use, use resources that are consistent with the value of the information produced by the model, use readily available input information, and instill user confidence and comfort. Sensitivity analyses should be conducted to learn the model, its behavior, and the key variables where most accurate input values are required.

SUGGESTED READINGS

Foster, G. R. 1982 *Modeling the Erosion Process*. In: Hydrologic Modeling of Small Watersheds. C. T. Hean H. P. Johnson, and D. L. Brakensiek (eds), American Society of Agricultural Engineers. St. Joseph, MI.

Haan, C. T., B. J. Barfield, and J. C. Hayes. 1994. Design Hydrology Sedimontology for Small Catchments. Academic Press, San Diego, CA.

6

Erosion Measurement

Our understanding of erosion processes is based on measurements taken by researchers using various measurement techniques for many years. Erosion measurements are the foundation of both water and wind erosion-prediction and erosion-control technologies, as discussed in Chapters 5 and 7. Erosion measurements and estimates provide the basis for government policies and programs, allocation of government resources, regulation of environmental impacts resulting from human activities, and conservation planning, as discussed in Chapter 8. In this chapter we examine (1) reasons for erosion measurement, (2) types of measurements, (3) erosion-measurement practices, and (4) evaluation of the data from erosion measurements.

REASONS TO MEASURE EROSION

Erosion is measured for three principal reasons (1) erosion inventories, (2) scientific erosion research, and (3) development and evaluation of erosion-control practices. Erosion inventories often consist of periodic erosion assessments for large areas using erosion-prediction technologies, such as the National Resources Inventory conducted at five-year intervals for private lands in the United States by the Natural Resources Conservation Service (USDA, NRCS, 1997b, revised 2000). These inventories are used in the development of public policies, as discussed in Chapter 8, and comparison of inventories through time can be used to evaluate the effectiveness of public policies. Erosion inventories also may involve direct measurement of erosion on small areas, such as representative erosion plots, hillslope profiles, or small watersheds. These measurements can be made

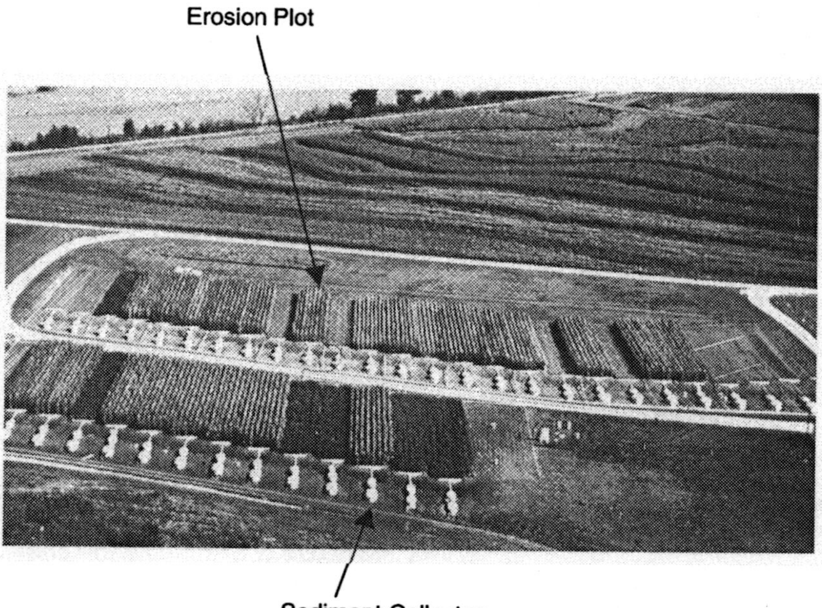

Figure 6.1 Erosion plots showing alternative management practices, Kingdom City, Missouri. Two-stage runoff and sediment collection at base of plots. (Courtesy of USDA, ARS.)

before land disturbance to establish baseline erosion conditions or after conservation or reclamation to determine the effectiveness of erosion-control practices (Toy, 1989). A single reconnaissance erosion inventory provides a general overview of (1) low- and high-erosion hazard sites based on past erosion damage, or (2) sites where severe erosion damage and high sediment yields are likely to occur if disturbed by human activity. These inventories are used to design erosion-control plans and allocate resources when detailed assessments of erosion are not feasible.

Erosion measurements are central to the scientific study of erosion. Studies of erosion mechanics, interrelationships among erosion processes, and relationships between erosion processes and environmental conditions are based on erosion measurements. The development of erosion-prediction technologies require erosion measurements. These data and derived erosion-prediction technologies are used to model landscape and landform development (e.g., Ahnert, 1988).

The development and evaluation of erosion-control practices rely on erosion measurements (Figure 6.1). Often, these data are used to rank practices based on their ability to reduce erosion rates for various climate, soil, and topographic conditions (Figure 6.1). During the 1980s, numerous studies compared the effects on erosion rates of conventional tillage, conservation tillage, and no-till practices.

TYPES OF EROSION MEASUREMENT

The types of erosion measurements differ for water and wind erosion. For water erosion, measurements can be made at various spatial scales because the sediment source-area boundaries can be established and runoff moves downslope in a predictable direction. For example, water erosion can be measured for interrill, rill, overland flow (interrill + rill) areas, concentrated-flow areas (channels larger than rills), fields, and small watersheds (overland flow + concentrated-flow areas, as well as sediment deposition), as described in Chapter 3. For wind erosion, the sediment source-area boundaries usually cannot be firmly established because wind changes direction, carrying sediment both into and out of the study area. Wind erosion is therefore measured by the sediment-transportation mode. Sediment is transported by saltation and creep and in suspension, as described in Chapter 3. It is difficult, and often unnecessary, to separate sediment transported by saltation and by creep.

Sediment deposition can be measured on hillslopes, in backwater areas, in drainage ditches and stream channels, on floodplains, in reservoirs, or in constructed sediment basins and terrace channels. Sediment characteristics can also be measured, such as particle-size distributions and bulk density. Where large amounts of sediment are deposited in small areas, such as in sediment basins, measurements can be quite accurate. Where small amounts of sediment are distributed over wide areas, such as on hillslopes, measurement is quite difficult.

Measurements also are made of the site characteristics that influence erosion and sedimentation, such as (1) weather and climate, (2) soil properties, (3) topography, including surface roughness, (4) vegetation, including canopy and ground cover, plant types and production levels, and (5) land management, including the types and intensities of land-disturbing activities. This information describes the environmental setting in which erosion and sedimentation occur and is required for selecting erosion-measurement techniques, "explaining" the variability of erosion rates in research, developing and applying erosion-prediction technologies, and designing conservation and reclamation plans.

Temporal Measurement Scales

Erosion measurements are made at various temporal and spatial scales. Practical temporal scales range from a single rainstorm or windstorm to several years. Erosion during a single storm may damage stream ecology or impair visibility along a highway. Storms of various recurrence intervals (return periods, Appendix B) are used to design erosion-and sediment-control practices. Erosion varies from season to season, and identification of high and low erosion-hazard periods can be used in scheduling land disturbances and installing erosion-control practices. Average annual erosion is often used in conservation and reclamation planning. Average annual erosion rates are compared to average annual soil-loss tolerances. Many years of weather data are necessary to estimate the recurrence intervals for storms, and several years of erosion measurements are necessary to determine seasonal or annual erosion rates depending on the overall variability of climate, soil, and vegetation conditions in the area. The effectiveness of erosion-control practices usually can be determined in just a few years.

Spatial Measurement Scales

Practical spatial scales for water-erosion measurements range from interrill and rill sediment sources on hillslopes to sediment discharge from watersheds. The presence of rills on hillslopes often is taken as evidence of erosion problems (Curtis et al., undated). The one-fourth or one-third of a field with the highest erosion rate commonly is used in conservation planning. Sediment discharge from watersheds is used in reservoir design. Whereas erosion control on uplands can control downstream sediment discharge, control of downstream sediment discharge alone does not preclude serious upland soil erosion and land degradation.

Spatial scales for wind-erosion measurements range from small plots or areas, such as a mill-tailing disposal site, to agricultural fields, to entire regions. Mill tailings are fine-textured, by-products of ore processing that frequently contain toxic materials. If allowed to dry, these materials are dispersed easily by wind erosion. Wind erosion frequently is underestimated as a cause of cropland topsoil depletion, as described in Chapter 1. Wind erosion from entire regions may contribute to atmospheric sediment loads. Sediment from the American Dust Bowl in the southern Great Plains reached Washington, DC on the east coast of the United States during the 1930s. As with water erosion, on-site wind-erosion controls reduce off-site environmental damage. To summarize, erosion measurements must be made at appropriate temporal and spatial scales, depending on the research problems and questions to be answered.

EROSION-MEASUREMENT PRACTICES

Every erosion-measurement project has a specific purpose, whether to inventory erosion rates, to study erosion processes, or to evaluate erosion-control practices. The quality of erosion measurements depends on the measurement plan and the measurement techniques. The measurement plan must produce data of sufficient quality and quantity to permit confident analyses and interpretations in addressing the research problems and questions. The measurement plan describes the required data, where measurements are to be made, the measurement techniques to be used, the steps to be followed in making the measurements, procedures for data cataloging and storage, the schedule of tasks, the methods to be used in data analyses and interpretation, and the personnel, time, and funding resources necessary to implement the plan. The measurement plan must be feasible under prevailing site conditions and with available resources. There are common and unique features of measurement plans for erosion inventories, scientific studies, and evaluations of erosion-control practices.

Erosion Inventories

Measurement plans for erosion inventories include (1) the procedures for selecting measurement sites, (2) the measurement frequency and duration at those sites, and (3) the measurement techniques for erosion processes and environmental conditions.

Site Selection

Erosion-measurement sites are selected according to a sampling strategy. Sampling alternatives include purposive, random, uniform, stratified, and complex sampling. *Purposive sampling* refers to site selection according to some purposive principle (Toebes and Ouryvaev, 1970; Young, 1972). This means that sites are deliberately (or purposively) selected as representative of the study area. For example, if the purpose of a study was to

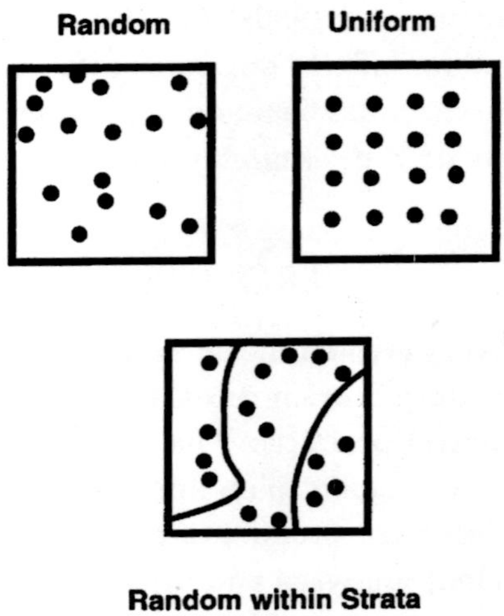

Figure 6.2 Random, uniform, and stratified (complex) sampling.

examine the relationship between erosion rates and soil characteristics in a particular area, the measurement sites would be chosen to include a range of representative soil characteristics for that area.

Purposive sampling is based on the professional judgment of the researcher. Selecting representative sites can be more difficult than assumed, however, as discovered by Griffin et al. (1988). In this study, uniform sampling produced a more representative site sample than purposive sampling. When resources constrain the sample size to a few sites, purposive sampling may be the only feasible alternative, although there is the possibility of bias in the resulting sample.

Common sampling strategies include random, uniform, and stratified sampling (Figure 6.2). From a statistical perspective, random samples are preferred because each point in the study area has an equal chance of inclusion in the sample, and the prospect of sampling bias is eliminated. There is no guarantee that random sampling will produce a set of study sites that completely represents the variability of environmental conditions within a study area or in the correct proportions.

With uniform sampling, a grid is placed over a map of the study area and measurement sites are located at the intersections of the grid lines. This method ensures uniform spatial coverage of the study area. If there is a periodic repetition of features or conditions within the study area, however, such as a nearly uniform spacing of stream channels, and the sampling interval happens to match this periodicity, the sample will not be representative of the study area. Once the spacing of the grid lines is chosen, each point in the study area no longer has an equal chance for inclusion in the sample and sample bias is a possibility.

With stratified sampling, the study area is divided into strata (areas sharing a common characteristic) based on a selected criterion, such as soils, topography, vegetation, or land use and management. Then measurement sites are selected randomly or otherwise within each strata. This procedure ensures that sites within each strata are represented in the sample. For example, if there are three soil types in the study area, all three types are represented in the sample. The number of measurement sites in each strata can be weighted by the proportion of the total study area contained within each strata.

Complex sampling strategies are possible that combine features of the methods described above. The National Resources Inventory for nonfederal lands in the United States (USDA, NRCS, 1997b, revised 2000) is based on more than 800,000 scientifically selected sample sites A two-stage stratified-area sampling procedure was developed for the entire United States. The first-stage sampling units are areas or segments of land designated as *primary sampling units* (PSUs). The second-stage sampling units are sites located within each PSU. Sample sites are located in all counties and parishes of all 50 states, in the District of Columbia, Puerto Rico, Virgin Islands, and selected portions of the Pacific Basin.

Statistical procedures are available for determining an adequate sample size (e.g., number of random points, spacing of grid lines). For example, the *best estimate of the standard deviation* for a sample can be used to compute the sample size necessary to establish the true mean of a variable with a probability of 95%, assuming a normal frequency (probability) distribution. When the standard deviation for the variable is large, the sample size also is large.

All sampling strategies possess advantages and disadvantages (Chorley, 1966; Richards, 1981; Young, 1972), and the "best" strategy is one tailored to the purpose of the erosion-measurement project, taking into account resource constraints. Practical considerations influence the final selection of measurement sites. Legal or physical obstacles may prevent site access. For example, a farmer may limit access through fields to fallow periods. A shepherd may not allow access to lambing pastures during the spring of the year. There may be no roads near measurement sites as well as intervening terrain that precludes overland travel. Access may vary seasonally; roadways may be unusable during the wet season. Electrical power may not be available near measurement sites.

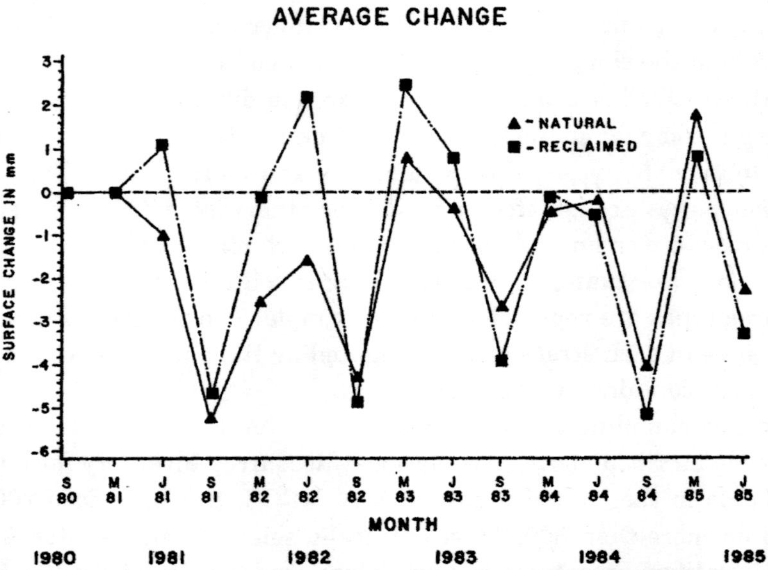

Figure 6.3 Seasonal changes in surface elevation, east-central Wyoming.

Measurement Frequency and Duration

Once the measurement sites are selected, the measurement frequency and duration are chosen. Measurement frequency refers to the number of times that measurements are taken during the measurement project. Measurements might be made several times during a runoff event, after each runoff event, daily, weekly, monthly, seasonally, or annually. Only a single observation may be needed to conclude that serious erosion is occurring in a particular area, while numerous measurements at very short intervals may be needed in studies of particle detachment by raindrop impact or abrasion of soil surfaces by windblown sediment. Frequency depends to some extent on whether measurements are made manually or by automated equipment.

Measurement duration refers to the length of time that erosion measurements are made at the selected sites. The duration should be sufficient in length to capture the temporal variability of erosion processes. Figure 6.3 shows the changes in surface elevation for a site in east-central Wyoming, due to seasonal changes in weather, surface-material properties, and erosion rates during a five-year period. Rainfall and runoff at particular sites can vary substantially, as documented in long-term records but not evident in short-term records (Knisel et al., 1979). If the purpose of a study is to determine average annual soil loss, fewer measurement years would be necessary in Mississippi, where large amounts of erosive rainfall occur each year, than in Arizona, where rainfall is infrequent and most erosion is caused by very few rainstorms each year. In addition to the scientific requirements of the data, measurement duration also is influenced by long-term (1) site availability, (2) research personnel availability, and (3) equipment availability, durability, and maintenance requirements.

Measuring Erosion and Site Conditions

As discussed in a subsequent section, measurement techniques such as erosion stakes, pins, bridges, frames, and the LEMI, can be used to inventory erosion. Measurements along hillslope profiles or across hillslope traverses integrate changes in erosion caused by spatial variabilities in soil, topography, and vegetation characteristics. These techniques are inexpensive, which often allows numerous sample sites to be measured at frequent intervals for several years, but these techniques also vary in measurement accuracy. In all cases, the equipment must be properly constructed, calibrated, installed, operated, and maintained, and equipment operators must be properly trained.

As noted previously, site characteristics also are measured, including (1) climate, (2) soil properties, (3) topography, (4) vegetative canopy and ground cover, and (5) land management. Meteorologists, soil scientists, hydrologists, engineers, geologists, geomorphologists, geographers, and botanists have developed specific equipment and procedures to measure most environmental variables. In this book we do not discuss these techniques; however, descriptions can be found in manuals and handbooks such as those listed in the Suggested Readings at the end of this chapter.

Scientific Erosion Research

Scientific research is conducted to improve our understanding of erosion. These studies require carefully constructed research designs, including detailed measurement plans and very accurate measurement techniques. Some of the factors that affect erosion usually are held constant while other factors vary so that the influence of the variables on erosion can be determined. A common research goal is development of cause-and-effect relationships among erosion processes and variables that influence these processes. These relationships often are expressed in equations whereby values of the dependent variable (e.g., erosion rate) can be predicted or estimated from values of the independent variable (e.g., slope steepness). Some studies examine individual erosion processes, such as detachment by raindrop impact, while other studies examine small process groups, such as interrill and rill processes together.

An essential element of individual-process studies is physical or mathematical isolation of the process to be examined. This necessitates a thorough understanding of erosion processes and the factors that influence them, as described in Chapter 4. For example, an experiment can be designed so that only raindrop impact is responsible for particle detachment and transport. To measure rill erosion, we could subtract interrill erosion measurements from total erosion (interrill + rill) measurements.

Field Measurements

Erosion research may be conducted in the field or in the laboratory. There are advantages and disadvantages to each approach. Field measurements are made under actual climate conditions and existing conditions of topography, soils, and vegetation, whether natural, managed, or reclaimed. Measurements on undisturbed sites can be compared with measurements on managed or reclaimed sites. There usually is the opportunity and space to measure erosion under varying conditions of soil types, vegetative covers, management, or slope aspect.

There also are disadvantages with field measurements. Some research control typically is sacrificed with field measurements. Prevailing climate conditions during short-term studies may not reflect the average climate conditions. High-intensity rainfall or wind storms may exceed equipment design capacities, resulting in lost records and equipment damage. On the other hand, only eight runoff events occurred during one five-year erosion study at one coal mine in Wyoming. Wind may blow simulated rainfall away from the research plot. Field-site surfaces and equipment may be disturbed by wildlife, domestic animals, or vandalism. Temperature, humidity, and windblown dust may affect the reliability of measurement equipment. In addition, there are expenses for travel, subsistence, and periodic equipment maintenance. Field conditions sometimes are uncomfortably hot, cold, or wet, not to mention the annoying insects and reptiles.

Laboratory Measurements

Two major advantages with laboratory measurements are the ability to carefully control experimental conditions and measure erosion processes. Both of which improve data accuracy (Figure 6.4). The range of values for each variable can be adjusted, within equipment constraints, to reflect the range of values observed in the field. Laboratory experiments often are short in duration (several minutes to several hours), so measurements can be taken at very short intervals using automated recording and data-logging equipment. Experiments are repeated to produce the data needed for reliable analyses, laboratory measurements usually are made at convenient times, at convenient locations, and under convenient conditions.

Laboratory measurements have their limitations and cannot be used to study some important factors affecting erosion. For example, the influence of soil and plant characteristics on erosion processes are not easily measured in laboratory experiments. Similarly, temporal changes in environmental conditions (e.g., climate, soils, vegetation) are not easily measured. Laboratory experiments commonly are scale-model studies because of the difference in size between experimental and field conditions. For example, a slope length of 20 ft (7 m) is relatively long for the laboratory but rela-

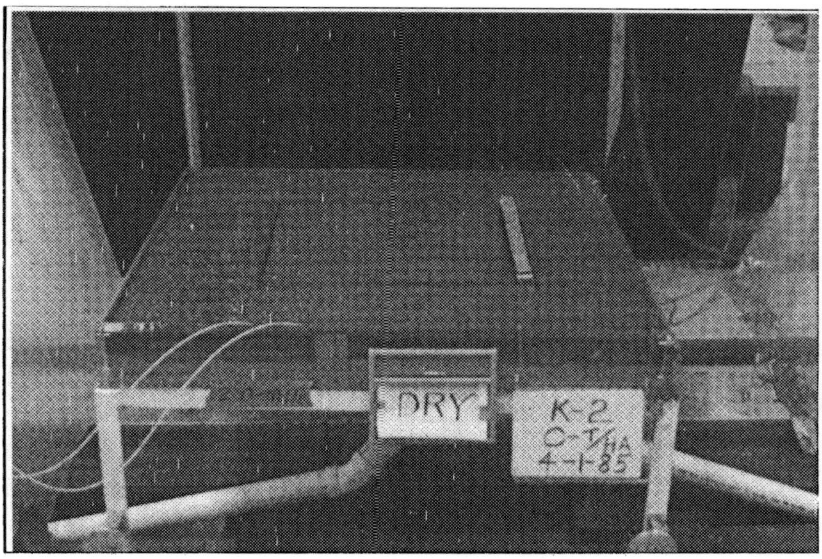

Figure 6.4 Laboratory interrill erosion experiment. (Courtesy of USDA, ARS.)

tively short for field conditions. As a result, there sometimes have been questions concerning the extent to which the laboratory measurements truly represent field conditions.

Evaluating Erosion-Control Practices

The performance of erosion-control practices in reducing erosion rates varies with climate, soil, topography, quality of installation, and frequency of maintenance as discussed in Chapters 7 and 8. In addition, practices vary in their cost and ease of installation and maintenance. Performance and cost generally determine the suitability and acceptability of practices for specific site conditions. Therefore, the performance of practices must be evaluated under various conditions and application intensities.

The performance of erosion-control practices is measured best under natural field conditions. However, the variabilities of soils and topography over short distances make it difficult to measure the specific effects of the practices alone on erosion rates. So performance usually is measured on experimental plots. Performance is measured best under natural rainfall conditions. However, 3 to 20 years of measurements may be needed for a thorough evaluation, depending on climate conditions. So performance often is measured using simulated rainfall. Uniform rainfall intensity allows a ranking of practices by performance rather than absolute measurement of a practice's ability to reduce erosion rates under field conditions.

Practices to control interrill and rill erosion on the overland-flow portions of landscapes typically are evaluated using experimental plots about 10 to 75 ft (3 to 25 m) in width and 35 to 75 ft (10 to 25 m) in length. Plots should be of sufficient width to represent the spatial variability of rows on cropland and shrubs on rangeland. Erosion plots are discussed later in this chapter.

A set of plots with the same soil characteristics and slope steepness are needed to compare selected erosion-control practices. At least three plot-set replications are needed to establish that one practice is superior to another in reducing erosion rates because of the variability in measured erosion, especially when erosion rates are low. Two plot-set replications may be sufficient to rank the practices if the difference in measured erosion between the practices is large. One plot surface treatment provides the reference against which other practices are compared. For example, the reference plot may be clean-tilled and fallow.

Agricultural Example

A comparison of practices might consist of three conservation tillage systems, where each is operated at three levels of crop-residue burial. For this study, 30 plots are required [3 replications (3 tillage systems × 3 burial levels) + 3 reference plots]. Finding an area where 30 plots can be located with the same soil and slope steepness is very difficult. The first priority is maintaining uniform soil characteristics; the second priority is maintaining uniform slope steepness. The hillslope profile must be straight because concavity may induce deposition that decreases the measured erosion. Erosion-prediction technologies can be used to adjust for the effects of differences in slope steepness if the steepness remains uniform along the plot.

Erosion measurements are made for each plot as the basis for the comparison. In addition, measurements of rainfall intensity, slope length and steepness, soil characteristics, and vegetation and crop residue are used in developing guidelines and specifications for erosion-control practices and in developing erosion-prediction technologies.

Although many practices for interrill and rill erosion control can be studied using plots, other practices, such as contouring, terracing, and vegetation strips are studied best using small, first-order, watersheds within farm fields. These watersheds typically are about 2–5 ac (1–2 ha) in size. Studies of ephemeral gully erosion and its control require larger watersheds of about 5–100 ac (2–40 ha) in size.

Wind Erosion

Measuring wind erosion requires different measurement plans and equipment. The wind arriving at the measurement area may already carry a sediment load. Both the inflow and outflow of sediment are measured. An increase in sediment reflects wind erosion; a decrease in sediment reflects deposition. Numerous measurement sites are established because all of the windblown sediment cannot be collected at a single point, such as the end of a plot or the mouth of a small watershed (Figure 6.5). Wind flows from various directions during the year and even during a storm. Sediment samplers should be oriented, or capable of orienting themselves, so that they are facing the wind direction. Finally, windblown sediment is transported at various heights above the surface. Therefore, measurements must be made at various heights in order to determine the vertical distribution of the sediment load.

Figure 6.5 Wind erosion of fields with removal of topsoil from hilltops. (Courtesy of USDA, NRCS.)

SELECTED MEASUREMENT TECHNIQUES

A measurement technique is the sequence of steps or operations that produces a value representing the magnitude of a selected variable. A measurement technique includes the equipment and the procedures for using the equipment. The choice of measurement techniques depends on the (1) erosion process or processes to be measured, (2) temporal and spatial scales of measurement, (3) uses of the collected data, and (4) available time, personnel, and financial resources. A complete survey of measurement techniques is not possible in this chapter. Some traditional, common, distinctive, or innovative techniques are selected to illustrate different approaches to measuring erosion. There are four fundamental ways to measure erosion (1) change in weight, (2) change in surface elevation, (3) change in channel cross-section dimensions, and (4) sediment collection from erosion plots and watersheds.

Change in Weight

Erosion measurement by a change in weight is based on the principle that erosion removes material from the source area. For example, a container of soil is weighed before and after an erosion event, and the change in weight is the erosion measurement. Some early studies of soil detachment and transport by raindrop impact used this technique. Generally, this measurement technique was used in laboratory or field experiments with small soil samples.

Change in Surface Elevation

Erosion measurement by the change in elevation is based on the principle that erosion and deposition by water or wind change the elevation of the land surface. Measurement of the surface elevation at two points in time indicates the net effect of erosion and deposition during that time interval. A lower elevation at the end of the time interval indicates erosion and soil-settling following disturbance. A higher elevation at the end of the time interval indicates deposition or lifting of the soil by frost heave or, perhaps, burrowing animals. These techniques are not well suited for use on newly disturbed agricultural, construction, or reclaimed lands where soil settling is occurring. To measure erosion and deposition, the effects of soil settling, frost heaving, or burrowing animals must be minimized. The linear (depth) measurement of erosion can be converted to a mass measurement by multiplying together the depth and soil bulk density.

Erosion Stakes and Pins

Stakes or pins are implanted in the soil and remain in place for the duration of the study. The pins should extend below the frost line in the soil so they are not lifted by frost heave. Sometimes, permanent or removable washers are placed over the pin to create a stable measurement platform on the soil that tends to average minor surface irregularities. The distance between the top of the stake or pin and the ground surface is measured at selected time intervals. The change in distance indicates the change in surface elevation: erosion or deposition (Schumm, 1964).

Erosion stakes and pins are rapid and inexpensive techniques for measuring changes in surface elevation. Although widely used, there are several serious error sources associated with the use of erosion stakes and pins, as discussed by Haigh (1977). For example, implanting the stake or pin disturbs the soil exactly at the point of measurement. The stake or pin disturbs water and wind flow across the soil surface. Erosion is accelerated around the stake or pin so that measurements indicate higher erosion rates than actually occur. Because of these error sources, erosion-pin measurements are biased to some extent. Unfortunately, it is not possible to determine the magnitude of the bias so that measurement adjustments can be made. Nevertheless, erosion pins have been widely used in erosion studies.

Erosion Bridges and Frames

Erosion bridges and frames overcome one of the major limitations of stakes and pins by moving the soil and flow disturbances away from the measurement points. Two permanent support rods are implanted in the soil and remain in place for the duration of the study. A cross-bar (bridge) is mounted and leveled between the two rods (Figure 6.6). Pins are inserted into predrilled holes in the cross-bar. The pins are gently lowered until they just touch, but do not penetrate, the soil surface. The distance between the top of each pin and cross-bar is measured to about 0.04 in. (1 mm). The length of the cross-bar and the spacing of the holes determine the number of measurements along the cross-bar and depend on surface

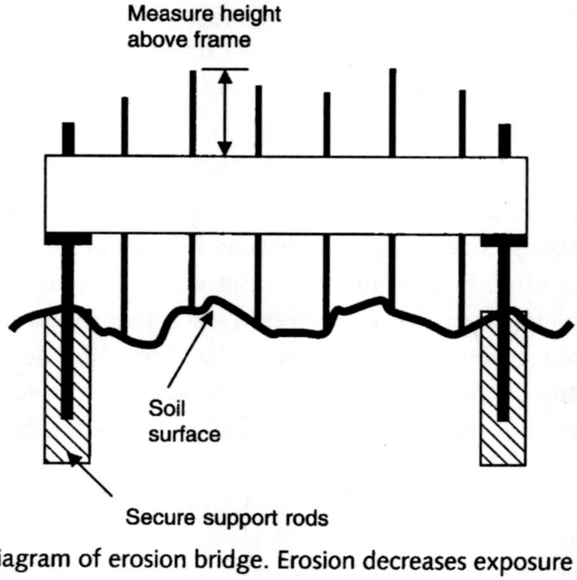

Figure 6.6 Diagram of erosion bridge. Erosion decreases exposure of pin above frame.

roughness and the data use. Erosion causes a decrease in the distance between the top of the pin and the cross-bar. Rill, profile, and microroughness meters are forms of erosion bridges.

Erosion bridges provide rapid and inexpensive measurements of changes in surface elevation (Hudson, 1965). Erosion bridges produce a spatially intensive set of measurements between the support rods. Disturbance of either support rod by animals or vandals probably results in the loss of all measurements for one time interval because the bridge cannot be repositioned exactly. The disturbed support rod is reset for measurement during subsequent intervals.

Erosion frames are based on the same concept as erosion bridges but produce a two-dimensional set of erosion measurements. Four permanent support rods are implanted in the soil and remain in place for the duration of the study. The erosion frame consists of connected erosion bridges, forming a matrix of holes with removable pins. One design consists of a 10.75-ft^2 (1-m^2) lattice of five bridges, with five holes in each bridge, for a total of 25 measurement pins (Campbell, 1970, 1974), although the number of measuring pins could be increased. Measurements are made in the same way as with erosion bridges.

Linear Elevation Measuring Instrument

A series of support rods are implanted along a hillslope profile. The linear elevation measuring instrument (LEMI) is constructed from two carpenters's levels set at right angles to each other with holes drilled and sleeved for measurement pins at either end of the larger carpenter's level. The LEMI is mounted and leveled on each support rod, measurement pins are lowered to just touch the soil surface, and the exposure of the measuring pins above the carpenter's level is recorded to the nearest 0.04 in. (1 mm) (Figure 6.7). The LEMI is then moved to the next downslope support rod.

LEMI measurements are spatially extensive, reflecting changes in elevation for paired and parallel hillslope profiles, on either side of the support rods, from the top to the bottom of the hillslope (Toy, 1983a,b). Measurements are taken at a distance of 1.64 ft (0.5 m) on either side of the support rod, so water and wind flow are not disturbed at the measurement point. The LEMI technique is somewhat more expensive per measurement because two people are required for the measurement process and more support rods are needed. LEMI measurements probably are slightly less accurate than erosion-bridge or frame measurements.

Figure 6.7 Linear erosion/elevation measuring instrument (LEMI). Measures change in surface elevation for two paired hillslope profiles on either side of support rod.

Photogrammetric Methods

Stereo photographs are taken of a hillslope surface with cameras at a known distance from the surface. The microtopography of the image area is analyzed using photogrammetric methods. This procedure is repeated at a later date and the change in the microtopography is computed. The use of stereo photography to measure the change in surface elevation requires precision equipment and accurate calibration. The photography platform may be on the ground and stationary or in aircraft flying at low altitude. A ground test on a roadcut indicated that one method was accurate to about ±0.12 to 0.20 in. (3 to 5 mm) (Cook and Valentine, 1979). A ground test on a streambank indicated that another method was accurate to about ±1.2 in. (30 mm) (Barker et al., 1997). Changes in surface elevation along a hillslope transect during a five-year period were measured using an aerial camera suspended 50 ft (15 m) above the terrain on a truck-mounted boom. This allowed construction of contour maps at a 1 : 24 scale with 0.5-ft (0.15-m) contour intervals (Piest et al., 1977). Concentrated-flow erosion (gully formation) was measured using aerial photographs taken from aircraft flying 590 to 902 ft (180 to 275 m) above the terrain, with consistent vertical accuracies of about ±1 in. (25 mm) or better (Thomas et al., 1986).

Photogrammetric methods offer a nearly noncontact method for measuring changes in surface elevation. It is still necessary to survey ground controls. These methods perform best where erosion rates are high, so the change in surface elevation is substantially greater than the resolution of the measurements. A very large point sample can be derived for the area depicted in the photographs. There are numerous options for analyses and display of the data obtained using photogrammetric methods. The available hardware and software supporting these methods improve continuously. The cost per measurement is small because a large number of point measurements are generated. The overall cost, however, is higher than for other methods because of equipment and analysis costs. The major limitation in using photogrammetric methods is that the soil surface must be photographed directly without vegetation or other obstructions. Bare-site photography produces bare-site erosion rates that are not representative of erosion rates on adjacent vegetated surfaces.

Laser Scanning of Topography

Optical laser scanning is another noncontact method used to measure the microtopography of a soil surface. This technique is used in conjunction with experimental field or laboratory erosion plots. In one configuration (Salvati et al., 2000), rails extend on either side of the erosion plot. A pair of line lasers and a digital camera are mounted on a traversing bar between the two rails (Figure 6.8). The lasers and camera move across the plot on the traversing bar. The lasers project a line of light onto the soil surface which is "sensed" and recorded by the camera. The digital data are calibrated to determine the surface elevation and cross-plot scan distance. A series of cross-plot microtopography profiles are constructed with a vertical resolution of about 0.02 in. (0.5 mm). The surface is scanned before and after an erosion event, and the change in elevation on the plot is measured.

Optical laser scanning of microtopography provides detailed and highly accurate measurements of changes in surface elevation due to erosion (Huang and Bradford, 1990; Römkens et al., 1988). This technology also is improving continuously. Again, the cost per measurement is small because a very large number of measurements is generated. The overall cost, however, is substantially higher than for several other methods.

Airborne laser altimeters also can be used to measure the topography

Figure 6.8 Laser scanner apparatus for measuring microtopography of surface. (Courtesy of USDA, ARS.)

and surface roughness properties of a landscape. For a study at the Walnut Gulch Watershed, Arizona, laser measurements were taken at horizontal intervals between 0.5 and 1 in. (0.0125 and 0.025 m) with a vertical resolution of 2 in. (0.05 m) from an aircraft flying at elevations between 490 and 980 ft (150 and 300 m) and at ground speeds between 164 and 328 ft/s (50 and 100 m/s). These measurements were used to quantify landscape topography, gully and stream channel cross sections, and vegetation properties. The cross sections can be used to estimate soil-loss volume from the gullies and stream channels (Ritchie et al., 1995). To measure changes in surface elevations for hillslope transects or channel cross sections using this technique, it is necessary to follow the exact flight line repeatedly.

Change in Channel Cross Sections

Channel erosion measurement by change in cross-sectional area is based on the principle that erosion and deposition processes change the dimensions of channels so that measurement of a channel cross section at two points in time indicates the net effect of the processes during that time interval. An increase in the cross-sectional area indicates net erosion, whereas a decrease in the cross-sectional area indicates net deposition. Somewhat different equipment is used for small and large channels.

Rills and Other Small Channels

The change in cross-sectional area of rills and small channels are measured with a rill, profile, or microroughness meter (McCool et al., 1981). These devices are based on the same principle as erosion bridges and frames, (Figure 6.6). With one design, 100 measuring pins spaced 0.4 in. (1 cm) apart produce a detailed diagram of the channel cross section (Simanton et al., 1978). Measurements at two time intervals can be analyzed manually, electronically, or photographically to determine the change in channel cross section. Several cross sections can be measured along the length of a rill to determine downstream changes in channel cross sections.

Large Channels

The changes in cross-sectional area of large channels can be measured using survey equipment. Permanent benchmark pins are implanted on either side of a channel to establish a survey line. Distance and elevation measurements are taken at intervals along the survey line (Figure 6.9). The surveys are repeated periodically to determine the changes in cross-sectional area (Toebes and Ouryvaev, 1970). Again, several cross sections can be measured along the channel length to determine downstream changes in cross sections.

Sometimes, rods or pins are placed in channels to measure bank erosion and bed scour, but the rods disturb the soil or alluvium and the water flow and cause erosion rates to appear higher than those that occur with undisturbed conditions. The maximum scour in a channel bed can be measured using *scour chains* wherein a length of chain is placed vertically in a hole drilled into the bed of a channel. The increment of chain laying horizontally after a runoff event in the channel indicates the maximum scour depth, and the sediment layer on top of the chain indicates the amount of sediment deposition following peak flow (Toebes and Ouryvaev, 1970). Some investigators prefer vertically placed ribbon (or bag twist-ties) that bend to the level of scour.

Sediment Collection

Several measurement techniques are based on the principle that water and wind erosion produce sediment, and this sediment can be collected to

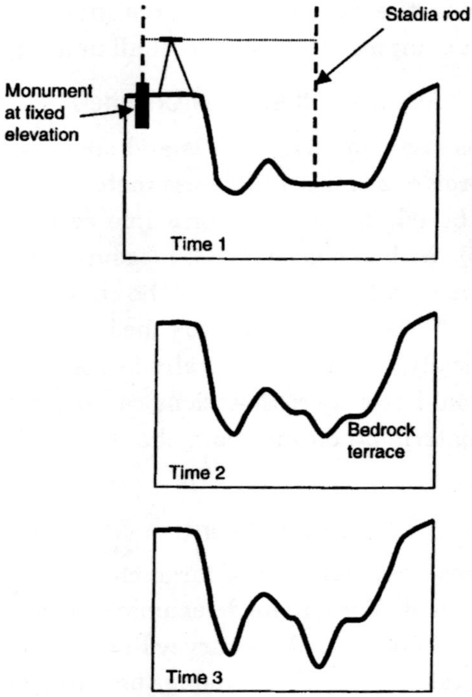

Figure 6.9 Channel survey to measure change in cross-section morphology.

measure erosion and deposition. Techniques based on this principle are used to measure (1) soil detachment by raindrop impact, (2) interrill erosion, (3) rill erosion, (4) total erosion, and (5) sediment yield from fields and small watersheds.

Raindrop Impact

The sediment eroded from interrill areas is detached by raindrop impact. The raindrop erosivity is a function of drop size and impact velocity. Techniques used to measure raindrop size and velocity include (1) paper blotter stains, (2) flour pellets, (3) optical methods, and (4) impact methods.

Paper Blotter Stains Raindrop size is computed from the size of the stain made on treated paper or slate. This technique works best for small drops where splatter is minimal. Experiments are conducted where drops of known size fall on the paper, the size of the stain is measured, and an equation is developed that computes the drop size from the stain size.

Flour Pellets Raindrop size is computed from the size of a flour pellet. This technique works best for large drops. Experiments are conducted where drops of known size fall onto a tray of flour to form pellets and the flour is heated to harden the pellets. The pellets are sieved from the remaining flour in the tray. An equation is developed to compute the drop diameter from the pellet weight.

The stain and pellet methods have been used widely because they are simple and inexpensive. However, measurement of a sufficient sample to characterize raindrop size distribution during a rainstorm is very time consuming. Automated methods to compute both raindrop size and fall velocity are replacing the stain and pellet methods.

Optical Devices High-speed photography, lasers, and infrared imagery are used to measure raindrop size and fall velocity. High-speed photographs of raindrops are taken during vertical fall using conventional high-speed cameras (Laws, 1941) or using high-speed line-scan videocameras (e.g., www.distrometer.at). The videocamera device, called a *distrometer*, measures rainfall intensity, drop size, vertical and horizontal velocity, and drop oblateness. An infrared optical device determines rainfall intensity, drop size, and fall velocity by measuring the variation in light intensity and the duration of the variation as a raindrop passes through the infrared beam (Salles and Poesen, 1999). Lasers are used to measure drop size and fall velocity based on movement of the drops through the laser beam (Kincaid et al., 1996).

Optical devices measure raindrop characteristics in different ways with somewhat different levels of accuracy. The equipment is expensive, compared to the stain and pellet techniques. Major advantages, however, are that these techniques produce very large raindrop samples and process the data automatically into various output formats.

Impact Methods The sound or pressure of raindrop impacts can be measured and used to determine the distribution of raindrop sizes. One instrument, called a *disdrometer*, transforms the mechanical momentum of a raindrop into an electrical pulse whose amplitude is a function of drop diameter (e.g., www.distromet.com). Again, the equipment is expensive but produces very large raindrop samples and processes the data automatically.

Rainsplash

With the traditional erosion concept, rainsplash was considered to be a discrete erosion process. Splash boards, funnels, and cups were developed to isolate and measure particle detachment and transportation by rainsplash. According to the data collected by these methods, rainsplash transported more soil downslope than upslope, resulting in net erosion. Other data, however, indicate that rainsplash accounts for only a small proportion of the soil transport from interrill areas to rills, and hence rainsplash is not an important erosion process (Meyer et al., 1975b). Most of the sediment is transported from interrill areas by interrill flow. Nevertheless, two rainsplash measurement techniques that have been used in the past are discussed below.

Splash Funnels A funnel is implanted in the soil and captures the sediment that is splashed into it (Bollinne, 1980). The rim of the funnel extends slightly above the surface to prevent runoff from entering the funnel. In some cases, a container is placed beneath the funnel to collect rainfall and sediment. The amount of splash erosion is determined by weighing the sediment in the funnel or container.

Splash Cups A tube is pushed into the soil and a larger collar or collecting tray is placed around the tube (Morgan, 1978). The collar is partitioned into upslope and downslope sections. A short wall surrounding the tray prevents runoff and exterior splash from entering the tray and captures particles propelled to the perimeter of the tray. The weight difference between the downslope side and the upslope sections of the tray is net downslope soil transportation, or erosion.

Although rainsplash does not appear to be an important sediment-transport process in soil erosion studies, the splash of contaminated sediment is a major transport pathway for radionuclides to reach plant surfaces. Additional research is needed to measure sediment splashed to various heights on plants by raindrop impact (Foster et al., 1985b).

Interrill Erosion

Erosion types within a spatial context are identified as interrill, rill, concentrated-flow area, and watershed erosion, according to the concepts presented in Chapters 3 and 4. Interrill erosion includes particle detachment by raindrop impact, some rainsplash directly to rills, and downslope sediment transport by surface flow. Studies of interrill erosion must be designed carefully to ensure that the measured sediment load contains only particles detached by raindrop impact. The measurement area cannot be so large that surface flow detaches particles.

Both laboratory and field experiments are used to study interrill erosion. For example, a 2 ft by 2 ft (0.61 by 0.61 m) soil pan is surrounded by a 1-ft (0.3-m)-wide apron of the same soil material and subjected to simulated rainfall. The soil apron is essential so that any soil splashed out of the pan is matched by soil splashed back into the pan. The soil pan is inclined and the unconcentrated, sediment-laden flow from the pan is withdrawn on the downslope side of the pan for laboratory analyses (Meyer et al., 1975b).

Interrill erosion on cultivated fields is measured using small plots, about 20 in. (50 cm) in width (depending on row width) and 30 in. (75 cm) in. length, installed across two facing row sideslopes (Figure 6.10). A

Figure 6.10 Field interrill erosion plot. (Courtesy of USDA, ARS.)

metal trough is positioned in the furrow between the sideslopes to collect and convey the runoff and sediment to a container at the end of the plot.

Rill Erosion

Again both laboratory and field experiments are used to study rill erosion (Figure 6.11). Rill plots may be 13 to 35 ft (4 to 10 m) or more in length with a preformed channel. A furrow/sideslope configuration, typical of tilled cropland, may be the best way to study rills (Meyer et al., 1975a). Simulated rainfall is applied to the plot. Sometimes supplemental flow is added at the upper end of the plot to simulate longer slope lengths. Runoff and sediment are measured at the downslope end of the plot. Flow velocity may be determined by measuring the movement of a dye injected into the flow.

When supplemental flow is added to the rill, the measured sediment represents erosion at the end of the simulated slope length rather than erosion for the entire slope length. Supplemental flow is added carefully at the upper end of the rill to minimize erosion at the point of entry. Similarly, the base level at the lower end of the plot is controlled carefully to prevent erosion or deposition. The rill should be sufficient in length that any erosion at the top and any erosion or deposition at the bottom accounts for a small proportion of the total sediment from the rill.

A part of the sediment measured at the end of the rill comes from in-

Figure 6.11 Rill erosion plot. (Courtesy of USDA, ARS.)

terrill areas adjacent to the rill. Small interrill plots are located near the rill and subjected to the same simulated rainfall. Rill erosion equals the total erosion measured at the end of the rill minus interrill erosion measured from the small plot.

Field and Small Watershed Erosion

The sediment transported from fields and watersheds includes interrill and rill erosion and may include erosion from concentrated-flow areas as well. For small study areas and low-to-moderate rainfall intensities, runoff and sediment may be captured for measurement in large tanks. Sometimes, however, the runoff is measured as it passes through a flume, and aliquot (fractional) samples of the runoff and sediment are taken at prescribed intervals by automatic samplers. Flumes are calibrated troughs through which runoff flows. As the discharge rate increases, the *stage* or depth of water in the flume also increases. A calibration equation for the flume is used to compute discharge from stage measurements. Various types of flumes and sampling equipment have been developed (Brakensiek et al., 1979; Toebes and Ouryvaev, 1970), but sediment deposition within the flume can be a serious problem affecting the stage measurements and discharge computations.

Multislot, rotating-wheel (Coshocton wheel), and traversing-slot (Figure 6.12) divisors split the flow into aliquot samples. For example, the erosion plots at Kingdom City, Missouri (Figure 6.1) are equipped with a two-step measurement system. When the volume of runoff and sediment reach the capacity of the first tank, a divisor directs one-third of the additional flow into a second tank. The remaining two-thirds of the runoff and sediment is discarded. Sediment samples are analyzed to determine the sediment concentration. The sediment concentration is multiplied by the runoff volume to estimate the total sediment mass for the event at the time of sampling. Detailed descriptions of the equipment to measure runoff and sediment from fields and watersheds is provided in the Suggested Readings at the end of the chapter. Acoustical methods are under development to measure the suspended sediment load in streams. (www.sedlab.olemiss.edu/cwp_unit/acoustic.html).

Aeolian Sediment Collection

As discussed previously, wind-erosion studies encounter additional challenges. Water moves in predictable downslope directions, whereas wind

Figure 6.12 Flume and sediment traversing slot sediment sampler at Walnut Gulch, Arizona. Note vertical traversing slot about one-third of distance from right side of flume. (Courtesy of USDA, ARS.)

changes directions seasonally and even during individual windstorms. Water and sediment sources are constrained by watershed divides or plot borders, whereas aeolian sediment sources may be local or distant. Water and sediment move downslope in flows of a few inches (centimeters) at most, whereas aeolian sediment may be distributed through several feet (meters) of a vertical air column. In addition, the wind entering a study area may already carry a sediment load.

Windblown sediment is measured in horizontal and vertical traps. Horizontal traps are used to collect sediment transported by creep and much of the sediment transported by saltation, as long as the box length is greater than the trajectory length of saltating particles. Horizontal traps consist of rectangular boxes embedded in the soil surface (Knott and Warren, 1981). The boxes should be positioned perpendicular to the expected wind direction or capable of rotating to that orientation.

Vertical traps or towers are used to measure sediment transported primarily by saltation but also to collect some sediment transported in suspension. Vertical traps and towers vary from simple slotted, vertical standing pipes to very sophisticated vertical towers. The backs of the slotted pipes may block airflow if not properly vented, resulting in very low sampling efficiencies. Wind-erosion measuring instruments are rated according to efficiency: the ratio of the sediment collected by the device to the amount of sediment actually in the airstream. The vertical towers have orifices at various elevations above the surface and are vented to minimize the differences in wind velocity and pressure between the inside and outside of the tower (Zobeck, 2001). A sampler that maintains the same wind speed through an orifice as the ambient wind speed at the same elevation is said to be *isokinetic.* Vertical traps and towers include mechanisms to rotate the orifices into the wind. The BSNE (Big Spring Number Eight, Big Spring, TX) is a popular vertical-tower sampler with an efficiency of about 90% for saltating particles (Figure 6.13). A tipping-bucket apparatus may be installed beneath the vertical tower to measure the accumulation of windblown sediment through time (Bauer and Namikas, 1998).

Suspended-sediment measurement requires isokinetic samplers. Samplers frequently consist of inlet tubes through which sediment is drawn to a filter by a suction source that is controlled to maintain the isokinetic condition. Instrument clusters are used to measure the sediment transported by creep, saltation, and suspension in a particular area (Fryrear et al., 1990) Recent innovations concerning wind erosion and its measurement are available on the USDA, ARS, Wind Erosion Research Unit Web site (http://weru.ksu.edu).

Erosion Plots

Erosion plots have been used in studies of erosion processes and in the development of erosion-control practices for several decades. Research designs based on erosion plots have proven to be effective and convenient for investigating a wide variety of erosion-research problems (Mutchler et al., 1994). When properly installed and maintained, erosion plots provide important experimental control, with certain factors that influence erosion processes serving as constants and other factors functioning as variables. Highly accurate measurement techniques, such as laser microtopography scanners, can be used to gather data on erosion-plot surface characteristics.

(a)

(b)

Figure 6.13 BSNE wind erosion sampler: (*a*) front view, (note dime for scale between first and second orifice); (*b*) side view, showing second sampler in background. (Courtesy of USDA, ARS.)

Erosion plots have been used in both laboratory and field research. Laboratory plots are carefully designed to achieve experimental control and maximum data accuracy, limited in design primarily by available financial resources. Field plots generally are located according to some sampling method. The specific site is chosen to include uniform slope steepness, soil type, vegetation characteristics, and slope aspect. Plots are usually placed in the straight middle segment of hillslopes so that erosion processes are not influenced by the lower concavity of the hillslope profile. Plots represent the upper end of the overland-flow slope length. Upper and lower plot borders are positioned on the contour and constructed of durable materials. Runoff and sediment collection equipment are installed along the downslope edge of the plot. For small plots, such as interrill plots, all the runoff and sediment may be captured in containers. For large plots, runoff and sediment are sampled using a multistage, multislot, rotating-wheel (Coshocton wheel) or traversing-slot (Figure 6.12) device, as described above.

Plot-surface preparation depends on the purpose of the research. For example, the vegetation may be retained on the surface of one plot while the vegetation is removed from an adjacent plot to test the effect of vegetation on erosion processes. One plot remains smooth while an adjacent plot is roughened to test the effect of microtopographic roughness on erosion processes. A standard treatment serves as the basis of comparison for other surface treatments.

Plot size also depends on the purpose of the research. Only interrill erosion is measured on the small [1 yd^2 (1 m^2)] plots. Rills do not develop on such short plots. Total erosion (interrill + rill erosion) is measured on larger plots (Figure 6.14), ranging in length from 36 to 650 ft (11 to 198 m) and ranging in width from 6 to 150 ft (2 to 46 m) (Brakensiek et al., 1979). These dimensions were chosen so that the resulting surface area was a fractional part of an acre. For example, a plot 72.6 ft (22 m) in length by 6 ft (1.8 m) in width is 0.01 acre (0.004 ha) in area. As described earlier, water and sediment can be added to the tops of small plots to simulate longer lengths. Sometimes, large plots several acres (hectares) in area are required in erosion studies (Toebes and Ouryvaev, 1970). The effects of some erosion-control practices, such as contouring and terracing, cannot be evaluated using small plots but require large plots or small watersheds.

Figure 6.14 Typical erosion research plot. (Courtesy of USDA, ARS.)

Rainfall Simulation

Rainfall simulators were developed because several years may be required to measure erosion resulting from natural rainfall. Rainfall simulators have been used in laboratory and field experiments to study the fundamental mechanics of runoff and erosion processes and to evaluate the effects on erosion of (1) tillage methods, crop type, crop rotations, residues, other soil-surface covers, and land management; (2) slope length, steepness, and shape; and (3) soil characteristics and erodibility. In addition, large rainfall simulators [1.5 acres (0.6 ha)] are used to demonstrate soil erosion and best management or other erosion-control practices (Dillaha et al., 1988). The erosion plot is divided into subsections with various treatments. The runoff and erosion rates from each subsection can be compared following the rainfall simulation.

Rainfall simulation was used to evaluate the success of surface-mine reclamation (Lusby and Toy, 1976). Erosion rates are useful indicators of reclamation success because they integrate the effects of soil, topography, surface cover, and land management. If reclamation practices have been successful, the runoff and erosion rates from reclaimed lands should be similar to the runoff and erosion rates from undisturbed lands. Clearly, rainfall simulators are important research tools, and in many cases, there is no feasible alternative to the use of rainfall simulators in erosion research (Meyer, 1994, Neff, 1979).

Simulating Natural Rainfall

A major issue concerning the use of a rainfall simulator is the extent to which the apparatus emulates natural rainfall characteristics. Ideally, rainfall simulators should effectively reproduce the rainfall intensity, raindrop-size distribution, raindrop-impact energy, and the spatial and temporal variabilities of natural rainstorms over the plot. Most simulators use nozzles to produce rainfall. One widely used nozzle produces a water-drop distribution that is close to, but slightly smaller than, that of natural rainfall (Bubenzer, 1979b). Consequently, the rainfall-impact energy from simulators using this nozzle is about 75% of natural rainfall. Another problem with this nozzle, and all other nozzles, is that producing an appropriate drop-size distribution results in a high rainfall intensity. Finally, the spray from these nozzles is fan-shaped. Therefore, the nozzles should be moved over the plot and operated intermittently.

Frequently, there are differences in the water quality of natural rainfall and the water used in rainfall simulations. Erosion scientists and engineers commonly ignore this potentially important variable because of the cost of obtaining a large supply of nearly pure water.

Types of Rainfall Simulators

The equipment used in rainfall-simulation experiments has evolved from the hand application of water to erosion plots with sprinkling cans at the Kansas Experiment Station in 1931 to computer-controlled devices for applying water, measuring runoff and erosion, and logging the data automatically (Foster et al., 1979; Laflen et al., 1991; Meyer, 1994 Renard, 1986; Simanton et al., 1991). Water has been applied at various intensities through pressurized nozzles, as described above, or by gravity-driven drop formers made of yarn, wire, glass, steel, brass, or polyethylene tubing. Detailed reviews of rainfall simulation are provided by Brakensiek et al. (1979), Bubenzer (1979a), Hudson (1981a), and Meyer (1994).

Three major rainfall-simulator designs have evolved since 1931. The first widely used simulator for evaluating erosion-control practices, known as the *rainulator*, produces raindrop characteristics similar to those of natural rainfall, with fan-jet nozzles (Meyer and McCune, 1958).

Figure 6.15 Rotating-boom rainfall simulator. (Courtesy of USDA, ARS.)

The nozzles are mounted on carriages that move back and forth across plots. This simulator employs a complex valve and electrical control system to operate the nozzles in a particular sequence for rainfall intensities of 1.25, 2.5, and 5 in. (3.2, 6.4, and 12.7 cm) per hour. A second design that is much simpler, less expensive, more portable, and nearly maintenance-free compared to the rainulator is the *rotating-boom* simulator (Figure 6.15). This simulator uses the same nozzles, flowing continuously, rotating above and across the plot (Simanton et al., 1991; Swanson, 1965).

A problem with these simulators is that rainfall intensity cannot easily be varied to match typical storm patterns for a particular area. Another problem with intermittent rainfall simulators, such as the rainulator and the rotating boom, is that local rainfall intensity is very high during the time that the nozzle passes over a particular location on the plot. An advantage with the rotating-boom simulator is that the interval between passes is less than with the rainulator, but still longer than desirable. The infiltration rate is partly a function of this local rainfall intensity and the interval between passes.

A third rainfall-simulator design oscillates the nozzles back and forth and can reduce the interval between passes from about 20 seconds for the rainulator to 1 second, improving rainfall and infiltration characteristics. This simulator is operated by a programmable controller so that rainfall intensity can be varied through time to represent natural rainfall patterns (Foster et al., 1979).

The characteristics that influence the choice of rainfall simulator for a particular research project include (1) natural rainfall replication, (2) cost, (3) ease of transport, set up and operation, and (4) maintenance requirements. All rainfall-simulator designs are compromises among these characteristics.

Sediment Analysis

Different erosion processes result in different sediment characteristics (Chapter 4). For example, with water erosion, coarse-textured sediments are deposited where the sediment load exceeds the transport capacity, while fine-textured sediments are carried farther downslope or downstream. Sediment samples are analyzed in the laboratory to determine physical properties, such as sediment concentration, particle-size distribution, and density. Chemical characteristics may be measured to determine the transport of pollutants. There are numerous laboratory manuals that describe the proper procedures for sediment analyses, for example, *Agricultural Handbook 224* (Brakensiek et al., 1979)

It is important, however, to emphasize one point. The standard practice for particle-size analysis of soils usually calls for the dispersal of samples into primary particles (Appendix A). Sediment, however, often is eroded and transported as both primary particles and aggregates, so the size and weight distribution of the sediment as it is transported is physically more meaningful than the percent of primary-particle sizes in the sample. In other words, for many erosion studies, the sample should not be dispersed prior to particle-size analyses. The difference between sediment size before and after dispersal is a measure of sediment aggregation.

EVALUATION OF EROSION MEASUREMENT

The ability to evaluate old and new erosion data can be very important. The extent to which measurements represent the temporal and spatial variabilities of erosion, together with measurement accuracy determine the value of the data for a particular purpose. Occasionally, a measurement appears to be erroneous or atypical, requiring a decision concerning use of the measurement. Erosion cannot be measured under all possible combinations of environmental conditions. Therefore, the available data often must be extrapolated to conditions not represented in the measurements. Data and their interpretations should be examined critically in relation to the purpose for which they were collected.

Temporal and Spatial Variability

The factors that control erosion vary temporally and spatially, as discussed in earlier chapters. Erosion measurements collected in the field or in the laboratory should reflect this variability. Erosion inventories, erosion research, and the development and evaluation of erosion-control practices based on a few measurements collected during short time periods, under particular environmental conditions, may lead to erroneous interpretations and conclusions because relationships among processes and controls may appear nearly linear within a restricted value range but curvilinear through a broader value range. Thus, measurements limited to a few years of low rainfall may misrepresent the real difference in the effectiveness among these practices, leading to costly mistakes in conservation planning.

Temporal Variability

Erosion varies seasonally because rainfall and wind erosivity, soil erodibility, soil crusting, and vegetation properties vary seasonally. An understanding of the seasonal variabilities and interactions among erosivity, erodibility, and vegetation is important in determining the effectiveness of cover-management systems and for conservation planning. For example, erosion can be substantially reduced by selecting a cover-management system that provides maximum cover during the period of highest erosivity. Consequently, erosion should be measured through all seasons except, perhaps, during those times when almost no runoff occurs or wind velocities are very low.

Erosion also varies from year to year. The preferred measurement duration for erosion and other environmental conditions depends on the climate variability in the area of interest. Again, it is the interaction of rainfall and wind erosivity, soil erodibility, soil crusting, and vegetation properties that is important. The year with the highest rainfall and wind erosivity may not produce the most erosion if much of the increased erosivity occurred while soil erodibility was low and vegetative cover was high. Conversely, a year with low rainfall and wind erosivity could produce considerable erosion if the period of high erosivity coincided with the period of high soil erodibility and low vegetative cover.

The long-term climate records of most places include a few high-energy rainstorms that were capable of causing severe erosion. The question arises concerning the appropriate influence of such rainstorms on conservation and reclamation planning. In some places, it appears that most erosion results from a few large storms (Edwards and Owens, 1991; Hjelmfelt et al., 1986; Larson et al., 1997). More than 30 years of measurements in many states, however, indicate that a rainfall factor used to estimate average annual soil loss must include the cumulative effects of the many moderate-sized rainstorms as well as the effects of the occasional severe rainstorms (Wischmeier and Smith, 1978). Unfortunately, it is not possible to predict when or if the high-magnitude, low-frequency event will occur during a particular period of interest. Recurrence intervals (return periods) are based on statistical approximations; it is difficult to predict the 100-year storm with only 25 years of data. A decision must be made, therefore, concerning the influence of such an event on the planning and design process. The decision usually is based on the consequences of plan or design failure.

Consideration of high-magnitude rainfall and wind events is necessary when designing erosion-measurement systems or erosion- and sediment-control practices for severely disturbed lands. High-magnitude events can damage or "wash out" equipment installations. At mine or construction sites, the on- and off-site consequences of high runoff, erosion, and sediment-discharge rates may be very serious. For example, a sediment-containment basin may fail, allowing high sediment discharges to enter a stream channel that destroys the aquatic habitat, contaminates a municipal water supply, and reduces the life expectancy of a reservoir. The 10-year, 25-year, 100-year, or even the probable maximum rainfall event may be used in project designs, again depending on the consequences of design failure.

Spatial Variability

Water erosion and deposition vary spatially, from upland erosion near watershed divides to sediment yield along a stream course. Upland erosion can reduce agricultural productivity and create severely rilled and gullied landscapes despite low downstream sediment yield because of deposition along the sediment-transport path between the site of erosion and the downstream site where sediment yield is measured. Conversely, high sediment yields may not be caused by high upland erosion rates but rather, erosion in concentrated-flow areas or stream channels. Neither upland erosion rates nor downstream sediment yields are necessarily good predictors of the other, as discussed in greater detail in Chapter 8.

Wind erosion varies across the length of a field, depending on the relationship between the sediment load and the sediment-transport capacity. One marked difference between water and wind erosion, however, is that a change in wind direction could return sediment to its source area. A particular field might be an erosion site one day and a deposition site the next day.

Measurement Accuracy

The research purpose typically determines the required measurement accuracy and hence the measurement plan and choice of measurement techniques. Generally, as the level of measurement accuracy increases, the required resources increase. An efficient measurement plan therefore is one that produces data of sufficient accuracy to serve the purpose, without wasting resources. A shortcoming of this approach is that future uses of the data cannot be anticipated. The essential question is whether or not the available resources are sufficient to produce the level of data accuracy necessary for a particular purpose.

Measurement accuracy can be discussed from various perspectives, utilizing various terminologies. A simple approach (Eisenhart, 1952) is used in this section to present the basic concepts. Measurement is a sequence of steps or operations that produce a value representing the magnitude of a selected entity, whether constant or variable. The accuracy of the measurement process reflects the degree of conformity between the values produced by the measurement process and the true value of that entity. An *accurate* measurement process is both precise and unbiased. *Precision* refers to the degree of conformity among the measured values themselves. Repeated measurement of the same entity by a precise process produces a set of very similar values (small standard deviation). *Bias* refers to systematic errors in the measured values. Repeated, biased measurement of the same entity results in a set of values having a mean (average) value

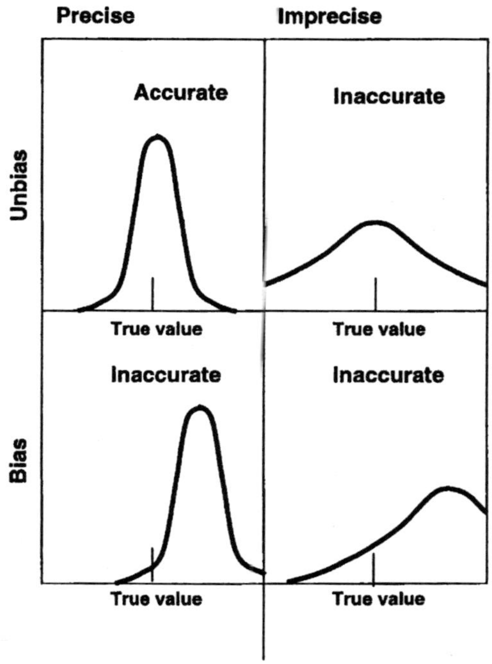

Figure 6.16 Measurement accuracy, precision, and bias. (Reproduced with permission, the American Society for Photogrammetry and Remote Sensing. Eisenhart, C., The reliability of measured values, Part I, fundamental concepts, June, 1952, 546.)

that is not centered on the true value but is somewhat higher or lower than the true value by the amount of bias.

As shown in Figure 6.16, there are four combinations of precision and bias with implications concerning measurement accuracy (1) precise and unbiased, hence accurate, (2) precise but biased, hence inaccurate, (3) imprecise but unbiased, hence inaccurate, and (4) imprecise and biased, hence inaccurate. The best choice is a measurement technique that is both precise and unbiased. Sometimes, however, this option is not available. One of the other combinations may be acceptable under certain circumstances. A measurement technique of high precision and known bias may be quite acceptable because the resulting values can be adjusted to remove the bias. But to know the magnitude of the bias, it is necessary to know the true value. A measurement technique with some imprecision, but unbiased, may be quite acceptable if the data set is large because the mean value of the data tends to center on the true value.

Bad Data

Most erosion data sets include a range of values, due to the temporal and spatial variabilities of environmental conditions. Sometimes a data set or value within a set appears nonrepresentative of erosion for particular environmental conditions. There are various reasons for values that seem out of place within the context of other data. An entire data set that differs from others collected under the same apparent conditions may be due to a faulty research design or inaccurate equipment calibration. A single value that differs from others may be due to equipment failure or a recording error.

Statistical approaches can be used to identify suspicious data. The *three standard deviations test*, for example, is based on the statistical probability that only 3 values in 1000 (1 in 333) are likely to differ from the value of the data set mean by ±3 standard deviations or more (Gregory, 1971). Frequently, however, recognizing nonrepresentative data is a matter of experience. Just as an experienced meteorologist can identify suspect data in precipitation records or an experienced hydrologist can identify suspect data in streamflow records, so an experienced erosion scientist can identify suspect soil-erosion records. There is no substitute for laboratory and field experience. When data errors are suspected, an attempt should be made to discover the cause before discarding data. The 100-year rainfall and erosion event may have occurred during your study. Then consideration should be given to the influence of the data or value on the research analyses and interpretations, as well as the decisions concerning conservation and reclamation planning. For example, consideration of the 100-year rainfall event might lead to the costly overdesign of practices intended to control average annual erosion, whereas ignoring the 25-year rainfall event might lead to the underdesign of sediment-containment basins, off-site environmental damages, violation of laws and regulations, and imposed fines or other penalties.

Data Extrapolation

Sometimes a research project to measure erosion is not possible because of time or other resource constraints. So existing erosion data are extrapolated to provide an erosion estimate. For example, erosion rates may have been measured for 20 to 80% vegetative cover, but we need to estimate an erosion rate for 10% or 90% vegetative cover. Statistical methods, such as simple regression analysis, can be used to compute an equation representing the average relationship between two variables. This equations can be used to estimate values beyond the range of measurements. Perhaps we need to estimate an erosion rate where the climate, soil characteristics, slope length and steepness, vegetative cover, and conservation practices all differ from those under which the available data were collected. Multiple regression analyses might be used to provide an erosion estimate, but erosion-prediction technologies, as discussed in Chapter 5, offer better procedures for data extrapolation because numerous interrelated variables influence erosion rates simultaneously, and these interrelations are incorporated into the recent erosion-prediction technologies.

Erosion data may be extrapolated through time. Environmental conditions are assumed to remain stable (stationary) during the period of extrapolation. For example, estimates of erosion rates for conservation and reclamation planning assume that climate conditions in the future will be about the same as those of the past. A climate record of 22 years or more is recommended to characterize climatic conditions for one erosion-prediction technology (Renard et al., 1997). Although climate, soils, and topography may remain stable into the future, land-use practices may change (e.g., conventional to conservation tillage), and such changes are difficult to anticipate.

Erosion data may also be extrapolated through space. The data collected from plots may be used to estimate erosion rates for hillslopes, watersheds, or even regions. Changes in erosion and deposition processes with spatial scale, discussed in Chapters 3 and 8, dictate that considerable care must be exercised in such extrapolations and the use of the resulting erosion estimates. For example, when using erosion-prediction technologies to estimate erosion rates for large areas, such as a watershed, the average watershed conditions should not be used as model inputs to calculate average watershed erosion rates. Rather, erosion rates should be calculated for several representative locations within the watershed, based on site-specific inputs, and the resulting erosion rates should be weighted by the area that those locations represent in determining an average erosion rate for the watershed. The validity of extrapolation depends on the purpose. Extrapolation procedures rest on assumptions and generalizations. The resulting erosion estimates may be quite useful for conservation planning but not useful in studies of erosion processes.

SUMMARY

Erosion research requires careful planning, accurate measurements, and thorough data analyses to arrive at the correct data interpretations and conclusions. The researcher is driven by the challenge of designing new research approaches and equipment, the excitement of breakthroughs to new knowledge, and the importance of expanding our understanding of erosion and erosion control.

SUGGESTED READINGS

Brakensiek, D. L., H. B. Osborn, and W. J. Rawls (coordinators). 1979. *Field Manual for Research in Agricultural Hydrology*. USDA Agricultural Handbook 224, U.S. Government Printing Office, Washington, DC.

Hudson, N. W. 1993. *Field Measurement of Soil Erosion and Runoff*. Food and Agriculture Organization of the United Nations, Rome.

Lal, R. (ed.). 1994. *Soil Erosion Research Methods*. 2nd ed. Soil and Water Conservation Society and St. Lucie Press, Ankeny, IA.

U.S. Department of Agriculture, Soil Conservation Service. 1972 (revised 1985). *National Engineering Handbook*, Sect. 4, *Hydrology*. U.S. Government Printing Office, Washington, DC.

U.S. Department of Agriculture, Soil Conservation Service. 1983. *National Soils Handbook*. No. 430. U.S. Government Printing Office, Washington, DC.

U.S. Geological Survey. 1997–1999. *National Field Manual for the Collection of Water-Quality Data*. USGS Techniques of Water-Resources Investigations, Book 9, Chaps. A1–A9, 2 vols. variously paged. U.S. Government Printing Office, Washington, DC.

7

Erosion and Sediment Control

Many land-disturbing activities such as farming, logging, mining, construction, waste burial, and military training expose land to the erosive forces of rainfall, runoff, and wind. Excessive erosion and sediment delivery occur if proper erosion and sediment control practices are not installed and maintained. Conservation is the selection of erosion and sediment control practices that provide the desired control while allowing the desired land use. Erosion-control practices are selected so that the on-site soil and land resource are protected. Erosion-control practices also are used to control sediment delivery. Sediment-control practices are applied specifically to control the amount, concentration, and size of sediment leaving the site. This sediment can cause downstream and downwind sedimentation as well as water and air quality degradation.

The principles discussed in Chapter 4 provide the basis for understanding erosion and sediment control practices and guiding their selection. In Chapter 5 we described erosion prediction technology that is widely used in conservation planning to ensure that practices are tailored to each specific site. In Chapter 6 we described how erosion and sediment delivery can be measured before and after installation of practices to determine their effectiveness. We also described how research is conducted to develop and evaluate erosion- and sediment-control practices. In the current chapter we describe erosion control principles that can be applied to any land use and provide examples of specific erosion- and sediment-control practices applied to particular land uses.

PRINCIPLES OF EROSION AND SEDIMENT CONTROL

The amount of sediment leaving any area is determined by either the amount of sediment made available by detachment or by the transport capacity of the erosive agents: runoff and wind. If the amount of sediment produced by detachment is less than the transport capacity, the amount of sediment leaving the area is controlled by detachment processes. In that situation, reductions in sediment delivery (yield) are most readily achieved by applying practices that reduce detachment. Controlling sediment delivery by reducing detachment has two benefits. In addition to controlling sediment delivery, controlling detachment protects the soil resource from degradation.

Sediment delivery can sometimes be controlled by reducing the sediment-transport capacity of the erosive agents. For this approach to work, the transport capacity must be reduced to a level less than the sediment load produced by detachment. Control of sediment delivery by reducing transport capacity always causes deposition, which enriches the sediment load in fine particles. The fine sediments are the portions of the sediment load that most seriously degrade air and water quality. Furthermore, the fine sediment often contains adsorbed chemicals that can be major pollutants. Therefore, when sediment delivery is controlled by reducing transport capacity, the reduction in the pollution potential of the sediment is less than the reduction in the sediment delivered. The coarse sediment particles cause greater downstream problems than do fine sediments.

The best way to control erosion, and thereby protect the soil resource, is by reducing detachment because control of erosion by reducing local transport capacity is a selective process where erosion removes fine particles and causes an increase in coarse particles in the soil. Thus, soil is degraded by erosion in two ways. First, erosion reduces soil depth. Second, when erosion is restricted by limiting transport capacity on the local scale, selective erosion enriches the soil in coarse particles, which often, but not always, degrades the soil by reducing both the moisture and nutrient-holding capacities.

Erosion- and sediment-control practices can be classified as cultural-management, supporting, and structural practices. Cultural-management practices typically are agronomic practices where vegetation and soil management are used to control erosion. For example, to control erosion, cultural-management practices manipulate vegetative canopy, ground cover from plant litter, applied ground cover materials, soil roughness, belowground biomass from live and dead roots and incorporated organic material, and mechanical soil disturbance.

Supporting practices are applied along with (hence, support) cultural-management practices. These practices often involve ridges and strips of vegetation oriented perpendicular to the direction of the runoff and wind. Terraces and diversions can be used to intercept runoff and reduce rill and concentrated-flow erosion. Subsurface drainage systems also reduce runoff and erosion. Both cultural-management and support practices are applied to the entire erosion-source area. Many of the erosion-control practices can be used to control both water and wind erosion, although adjustments are required depending on the type of erosion.

Structural practices typically are located at specific points to control erosion by channel-bed scour, headcuts in a permanent gully, or erosion on the outside bank of a stream-meander bend. Structural practices also can be barriers along the perimeter of the erosion source area, such as a windbreak to control erosion within the source area, a fabric fence, or a sediment-control basin to control sediment delivery from a construction site.

FUNDAMENTAL EROSION-CONTROL PRINCIPLES

Several basic concepts are used to control erosion on any land use, whether cropland, rangeland, construction sites, reclaimed land, or landfills. These principles are:

1. Maintain vegetative cover. Vegetative cover reduces erosion by providing canopy, plant litter for ground cover and incorporation into the soil, and root network. The goal is simply to maintain the highest-producing vegetation with the densest root network near the soil surface. Short vegetation is best for control of water erosion, and tall vegetation is best for control of wind erosion. Certain crop-production systems include row crops such as corn and soybeans that result in high erosion rates. A combination of these crops with dense hay crops in a crop rotation can be effective in reducing the overall average annual erosion.
2. Maintain ground cover. Keeping the ground cover by leaving

(*continues*)

(*continued*)

last year's crop residue on the surface, growing plants that produce a high level of litter, or applying cover, such as natural or manufactured mulch materials when no residues are available from plant growth, is highly effective for controlling erosion.

3. Maintain cover during periods when erosivity is highest if cover cannot be maintained at all times during the year. Having cover during the critical period is determined by the crop choice in a farming system or by scheduling soil disturbance for times other than the period of maximum erosivity.
4. Incorporate biomass into the soil. Adding to and incorporating organic material such as manure, sewage sludge (biosolids), or papermill waste in the soil can reduce erosion significantly.
5. Minimize soil disturbance, but if the soil must be disturbed mechanically, leave the soil surface rough, with large clods.
6. Add soil amendments such as PAMs (polyacrylamides) that reduce the erodibility of the soil and increase infiltration. Wind erosion and dust at a construction site can be controlled by water applications.
7. Add supporting practices, such as creating ridges and orienting them perpendicular to the direction of wind and runoff. Also, reduce surface runoff with surface drainage systems.
8. Prevent excessive rill erosion by avoiding long, steep slopes and water convergence. Terraces, diversions, strips of dense vegetation, and armored waterways for runoff disposal are effective in controlling this type of erosion.
9. Where possible, modify the topography, especially to avoid convex segments at the end of hillslope profiles. Use concave segments with very flat slope steepness at the end to induce deposition, which controls sediment delivery.
10. Avoid long field lengths, which can greatly increase wind erosion.
11. Use barriers such as windbreaks around fields.
12. Although the main principle is that sediment delivery from water erosion is controlled by reducing erosion on the source area, barriers can be placed around the source area to control sediment delivery when erosion control is not sufficient.

EXAMPLES OF WATER-EROSION-CONTROL PRACTICES

Many different practices are used to control erosion (Morgan, 1986). Our purpose here is to illustrate the main practices in a general way. As we discuss in Chapter 8, the choice of an erosion-control practice involves many considerations in addition to erosion control. The following discussion emphasizes the erosion-control principles, as summarized in the shadow box on pages 201 and 202, without considering profitability, convenience, and sustainability.

Cropland

Erosion differs among crops depending on crop type, how the crop is farmed, and the seasonal growing pattern of the crop in relation to the temporal distribution of rainfall/runoff erosivity at a location. Many complex and interacting variables affect erosion so that the relative effectiveness of an erosion-control practice varies with location (Table 7.1). The values for relative erosion-control effectiveness are the ratios of soil loss with the practice to the soil loss from a clean-tilled fallow condition as computed with version 2 of the revised universal soil loss equation (RUSLE2) (www.RUSLE2.com).

Choice of Crop

Table 7.1 shows how the relative erosion-control effectiveness varies by location because of the monthly temperature and precipitation and temporal variation of erosivity for that location. Notice that the order for corn by location is not the same as for wheat. A switch from wheat to corn at Boston, would result in a decrease in erosion, whereas a switch from wheat to corn at Madison would result in an erosion increase (Figure 7.1). Therefore, one way to reduce erosion on cropland is to choose a crop where the pattern of growth provides the maximum cover during the period of maximum erosivity.

Crop Rotation

Crops such as hay that produce a dense, short canopy with high amounts of root biomass reduce soil erosion. For example, the relative erosion-control effectiveness of a hay crop is 0.028, which is much lower than the 0.19 for the corn crop at Madison. When an erodible crop such as corn

Table 7.1 Relative Erosion-Control Effectiveness of Clean-Tilled Corn and Wheat for Erosion Control

Crop	*Location*	*Relative Erosion-Control Effectiveness*
Corn	Madison, WI	0.19
Corn	Memphis, TN	0.22
Corn	Boston, MA	0.15
Wheat	Madison, WI	0.14
Wheat	Memphis, TN	0.26
Wheat	Boston, MA	0.22

must be grown, it can be combined with a crop such as hay in a crop rotation to provide an overall relatively low erosion-control effectiveness of 0.11. The number of years of hay in the rotation can be increased or decreased, depending on how long the erosion-control effectiveness of hay is required to achieve the soil-loss tolerance or other conservation objective.

Conservation Tillage

In general, conservation tillage is a form of farming that leaves a portion of the residue from last year's crop on the soil surface. Clean tillage results in the burial of organic material of last year's residue at planting. Table

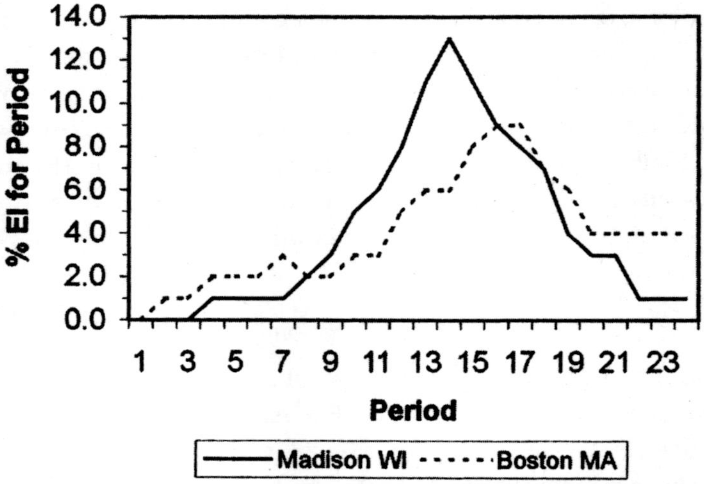

Figure 7.1 Erosivity variability for 24 periods annually at Madison, Wisconsin, and Boston, Massachusetts.

Table 7.2 Effect of Tillage Systems on Water-Erosion-Control Effectiveness for Corn at Madison, Wisconsin

Tillage System	*Erosion Control Effectiveness*	*% Residue Cover at Planting*
Clean till	0.19	5
Reduced till	0.10	32
No-till	0.01	76

7.2 shows values for erosion-control effectiveness of three types of conservation tillage systems with corn at Madison, Wisconsin. The erosion rate decreases as the amount of residue left on the soil surface increases. Leaving increased residue on the soil surface increases the erosion-control effectiveness of both the reduced till and the no-till practices, but residue cover is not the only factor affecting soil loss. The lack of soil disturbance and the accumulation of a high level of organic matter in the upper 2 in. (50 mm) of soil with the no-till practice greatly increases its erosion-control effectiveness.

Adding Organic Matter

Building the organic matter in the soil is a very important soil-management practice that increases infiltration and reduces erosion in conjunction with increasing crop yield. No-till practices increase organic matter at the soil surface, but organic matter can also be added to the soil with manure or sewage sludge applications. For example, adding a high amount of animal manure improves the erosion-control effectiveness from 0.19 to 0.13 for the clean-till tillage system. Growing cover crops during the winter and adding fertilizers to increase yield are other ways of increasing the amount of organic matter to reduce erosion.

Supporting Practices

High ridges oriented perpendicular to the runoff and wind direction reduces soil loss by about 30 to 50%, depending on several factors, including height of the ridges and rainfall/runoff erosivity. This support practice generally is known as *contouring*. However, contouring must be used with care. If the slope lengths are too long, the accumulation of runoff can cause the contouring to fail where runoff overtops a ridge. The location along the overland-flow slope length where contouring fails is known as the *critical slope length*. The cropping system can be changed to leave more residue on the soil surface and to leave a rougher surface, which increases the critical slope length. Terraces also can be installed on slopes to intercept runoff so that the slope length between terraces is less than the critical slope length. Narrow strips of grass (buffer strips) also can be installed at intervals along the slope to increase the critical slope length.

Chemical Amendments

PAMs (polyacrylamides) (Shainberg, 1992) added to the water introduced to fields for surface irrigation reduces soil erodibility and increase infiltration, to decrease the erosion associated with the surface irrigation.

Grazing Lands

The two main types of grazing lands are rangelands and pasturelands (Austin, 1981). Most rangelands are in the arid and semiarid regions in the western United States and much of this land is publicly owned. Also, management on these lands often is from an ecological perspective, purchased inputs (e.g., fertilizers) are minimal, and management is typically oriented toward maintaining native vegetation. Most pastureland is located in the eastern United States where rainfall is sufficient to make intensive management profitable, such as growing high-yield forage crops, fertilizing, and periodic pasture renovation and seeding.

Generally, the soil quality of both rangeland and pastureland is lower than for cropland. Also, the management options on grazed lands are much more limited than on cropland. Vegetative cover, including canopy, litter, and root biomass, are the key factors to be managed on these lands to control erosion. On rangelands, the principal management practice is grazing control. Several grazing practices are available. With one grazing plan, animals are on the land continuously at a sufficiently low density to avoid over grazing. For another plan, the land is grazed intensively for brief periods and then allowed to recover before animals are reintroduced. The erosion-control objective is to manage the animals so that vegetative cover is greatest during the season when erosivity is greatest.

Some lands, especially in the southwestern United States, have a rock "pavement" that develops over many years. This rock cover effectively reduces erosion just as plant litter reduces erosion. In other areas of the United States, rock cover is used specifically for erosion control (e.g., Box, 1981; Meyer et al., 1972). The practice is common on steep highway slopes.

Some rangelands are renovated periodically to establish desired plant communities and to eliminate undesired shrubs. Surface roughness left by renovation can increase infiltration and reduce runoff and erosion significantly for many years following treatment. Two of the major drawbacks to renovation in arid and semiarid environments are high cost and low economic return.

In addition to grazing, pasturelands are mowed for hay. Just as with rangelands, avoiding overly intense grazing and haying is critical to erosion control on pastureland. Maintaining vegetative cover is especially essential on these lands because rainfall and runoff erosivity can be high. Many pasturelands can be profitably renovated and fertilized at frequent intervals to maintain vegetative growth. Protecting pasturelands from gully erosion is especially important because gullies interfere with operations on pasturelands. Without cultivation to fill small channels, gully growth continues unchecked on pasturelands. Land use can move back and forth between cropland and pastureland, depending on the current economic farming conditions. Also, lands that are marginal as pasturelands can become forestland, by choice or happenstance. In the eastern United States, trees can invade pasturelands if volunteer growth of seedlings is not controlled and the pasture area will convert to an unmanaged woodlot.

Disturbed Forestlands

The litter cover on most undisturbed forests prevents excessive erosion (Dissmeyer and Foster, 1980). However, erosion is a problem where logging and grazing disturb the litter layer and expose “mineral” soil. The United States has three major types of forests, including (1) forests on public lands (e.g., National and State Forests), (2) commercial forests on both rented and privately owned lands, and (3) essentially unmanaged forests on private lands that often are grazed. Much of the publically owned forests are in the western United States on steep mountainous lands. Much of the commercial forest is in the southeastern on relatively flat but poor-quality land that was farmed around 1900. The other forestland use is farm woodlots, where the land is in forest because it is not suitable for either pasture or cropland. Five major types of forestland disturbance cause problems including logging operations, road building, and site treatment associated with seeding, fire, and grazing (Dissmeyer and Foster, 1981).

Logs are cut, trimmed, moved, and loaded onto trucks by machines. These machines disturb the litter and soil in varying degrees, depending on the intensity of activity in a particular area. Thus, the major erosion-control practices are intended to minimize the mechanical disturbance so that a minimal amount of surface cover is lost and to minimize the mechanical mixing of the soil. In this scenario, the high amount of organic material at the soil surface is buried in the soil. The disturbance can be substantial along skid trails, where logs are pulled, and landing areas, where logs are loaded. The primary erosion control practice is to minimize the traffic and intensity of activities in a particular area. Another erosion-control practice is to make sure that disturbed areas, especially skid trails, do not intersect with concentrated-flow areas where sediment is easily transported downstream. Still another practice is to leave a filter strip between disturbed areas and concentrated-flow areas so that sediment will be trapped as runoff passes through.

Roads create a special problem in forest areas because the drainage systems associated with roads often intersect stream channels. Roads often involve cuts and fills in steep terrain. High erosion rates can occur if these surfaces are not well vegetated. An important erosion-control practice for roads in forests is to establish rapidly and maintain vigorous vegetation on cut and fill surfaces. Mulch can be applied immediately after construction to provide immediate erosion control. Runoff from fill slopes should be directed to undisturbed areas so that runoff can infiltrate and the sediment is filtered from the runoff before it reaches a concentrated-flow area. A berm is often needed on the downslope side of the road to divert runoff to a stream channel with erosion control. Frequent diversions are needed along the road to direct runoff into undisturbed areas where concentrated-flow erosion is not a problem. The road surface itself can be a major erosion area and sediment source. Gravel on the road surface can greatly reduce erosion, but erosion rates can be very high following road maintenance until the surface becomes armored again.

Litter, slash, and other debris left by logging can hinder reseeding (planting seedlings to replace the harvested forest). Site preparation techniques including (1) raking and windrowing, (2) burning, (3) tillage, and (4) chopping to break limbs into smaller pieces and reduce excess material (Dissmeyer and Foster, 1980). Erosion increases as ground cover is reduced, soil exposed, and the organic material that was originally just below the soil surface is lost or buried deeper in the soil. Many of the nutrients in forest soils in this surface layer are occupied by fine tree roots, and removing them can greatly reduce the productive capacity of the soil. Site preparation should be planned and managed so that erosion is controlled to an acceptable rate. Erosion prediction technology can be used to select site preparation techniques, much like the selection of tillage practices for cropland (Dissmeyer and Foster, 1981).

The recovery period before the new forest develops a litter layer can be a period of excessive erosion which requires planting of temporary vegetation. Volunteer vegetation invades rapidly in most humid areas, so that sufficient erosion control is realized within five years or less, but the recovery rate depends on soil and rainfall conditions, requiring more time for arid conditions and poor soils.

Intense fire that removes both the litter layer and the root biomass immediately below the soil surface can result in very severe erosion. Also, fire can greatly reduce infiltration and increase runoff under some conditions (DeBano, 1981). Where possible, the intensity of the fire, such as in controlled burns, should be managed to avoid loosening surface cover and the underlying organic material. Revegetation is a standard erosion-control practice following fire.

Grazing in farm woodlots often results in loss of the litter layer, leaving the potential for significant increases in erosion. Although the litter layer may be significantly degraded, annual leaffall still provides ground cover, and the root network immediately below the soil surface is still present to significantly reduce but not eliminate erosion. Grazing should be managed, if not eliminated, on woodlots to maintain ground cover. Grazing also has the effect of significantly decreasing infiltration and increasing surface runoff, and in turn increases erosion because of absence of the litter layer.

Construction Sites

Erosion on construction sites can hardly be avoided because the very nature of construction exposes the soil. Because increased erosion is almost a certainty, the first step is to install sediment-control measures to prevent sediment from leaving the site. The fundamental erosion-control principle on construction sites is to minimize the soil area exposed at any one time. Avoiding soil disturbance during high-erosivity times of the year is mandatory. Thus, the construction is carried out in phases. As soon as one phase is completed, vegetation is established. If vegetation cannot be established immediately, a mulch cover using blown straw mulch or a manufactured (roll) material is applied. A critical factor in applying these materials is to ensure quality application. Very serious erosion can occur beneath the materials if they are not applied properly. A basic requirement is that mulch materials, whether blown on the soil surface or rolled on as a blanket material, must be anchored to the soil and provide continuous contact with the soil beneath. Otherwise, runoff can flow under the mulch and cause serious rill erosion. Crimping of straw mulch to anchor it into the soil often is needed to prevent runoff and wind from removing the mulch and exposing the bare soil.

A well-planned water-management system is critical to avoid excessive rill and gully erosion, which can easily occur on the bare soils and steep slopes, characteristic of construction sites. Diversions are frequently placed at the top of steep slopes to intercept runoff and to minimize run on downslopes vulnerable to erosion. Temporary channels can be protected with blanket material or straw bales placed across a channel during construction.

Figure 7.2 illustrates a variety of erosion-control practices and structures that are used for erosion control and shows structural practices to control runoff and to reduce sediment transport. Establishing permanent vegetation as quickly as possible after a site disturbance is needed to ensure erosion control. Removing, stockpiling, and replacing topsoil during site development can greatly facilitate establishing permanent vege-

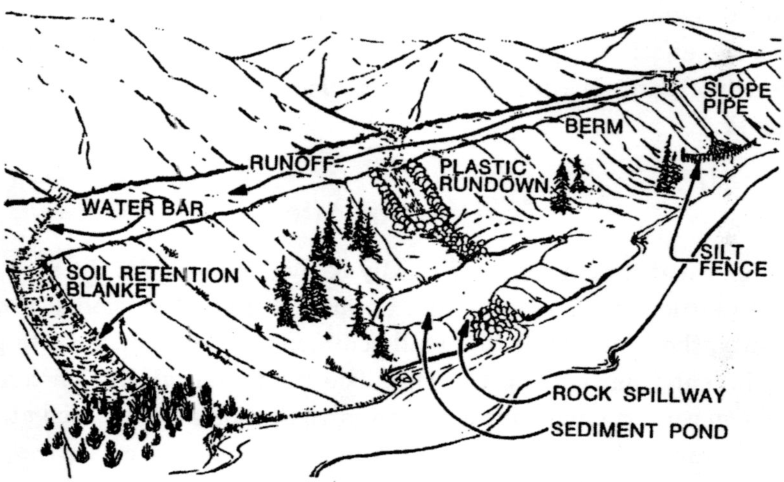

Figure 7.2 Erosion-control systems. (From Colorado Department of Highways, 1978.)

tation in a timely fashion. Failure to control erosion may require expensive reworking of the site.

Reclamation of Mined Land

Surface-mining operations totally change the landscape. For surface mining of coal, another regulatory requirement is that the land be restored to its approximate original contour and that if the land was prime farmland before mining, it must be returned to farmland. Reclaiming mine lands requires highly effective erosion-control practices to avoid severe erosion (Toy and Daniels, 1998). For surface mining of coal, state and federal regulations require that an erosion-control system using permanent vegetation be established before the security bond is released.

The requirement that the land be returned to the approximate original contour is especially challenging because landscapes evolve as determined by geology–climate–soil–vegetation–cover conditions. The reclaimed soil and cover conditions after reclamation almost always are very different from those before mining (Figure 7.3); note the difference in vegetation between the foreground and background in this photograph. The potential for erosion increases because of changes in infiltration, runoff, soil, and vegetation. The first step in erosion control on reclaimed land is to develop an erosion control plan. Although the land must be returned to approxi-

Figure 7.3 Erosion-control planting on a reclamation site in Wyoming.

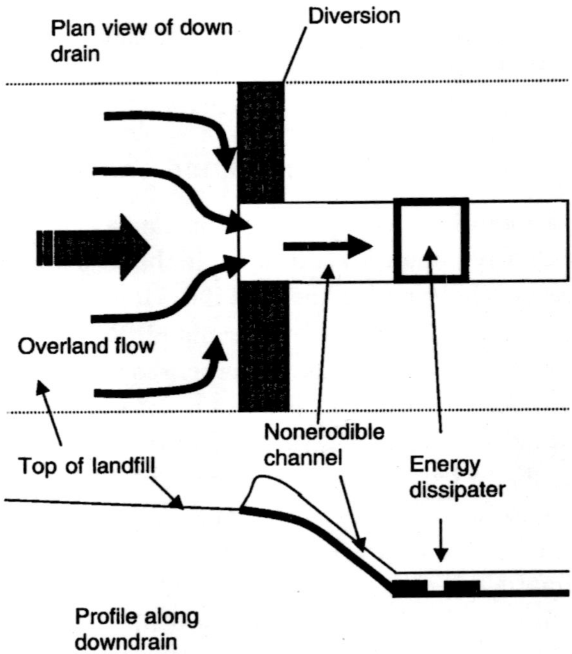

Figure 7.4 Structure for energy dissipation.

mate original contour, the reclamation planner has considerable flexibility in the design of the landscape. A key feature of the new landscape is the drainage system of channels. These channels must be designed properly to avoid gully erosion (Figure 7.4) and to avoid excessive long slope lengths on the overland flow areas (long slopes concentrate runoff and accelerate erosion). The drainage system is designed as part of the plan view of the landscape. Hollows in the plan view become areas of concentrated flow (Chapters 2 and 3).

Hillslope profile shape is another important design element (Toy and Black, 2000). Convex-shaped slopes (Figure 2.4*a*) must be avoided because of their high erosion potential. Complex slopes with a dominant lower concave shape induce deposition and help to control sediment yield. Also, this slope shape can help minimize the advance of gullies from concentrated-flow areas into the sides of hillslopes. In some cases, slope lengths can be excessively long, which could cause excessive rill erosion and ultimately lead to gully erosion. Although erosion control should be sufficient to prevent rill erosion, gully erosion definitely must be avoided. In fact, the security bond for a reclaimed site often will not be released if either excessive rill or gully erosion is present. A major factor in the final topography will be the cost of earthmoving, which is the major reclamation cost.

A key to successful reclamation is immediate erosion control. The topsoil stockpiled before mining is replaced on the graded "spoil" surface. Compaction should be alleviated in the upper layer of spoil material before the stockpiled topsoil is replaced. Compaction can greatly restrict infiltration and increase runoff and erosion. Ripping may be required to break up compacted soil. Standard agricultural methods are used to loosen the topsoil and prepare the seedbed. The topsoil provides a far better soil for establishing and supporting vegetation than does the spoil material. Even though topsoil has been replaced, amendments including fertilizer, lime, and other soil nutrients are applied as needed before seeding. The seed mix is selected in accordance with postreclamation land use.

Mulch commonly is applied to control erosion before vegetation becomes established, and fast-growing temporary vegetation may be grown until the permanent vegetation becomes established. Supporting practices of contouring and surface roughening to capture rainfall and runoff can help greatly in establishing vegetation and controlling erosion. The benefits of these supporting practices can persist for a decade or more in low-rainfall areas. The vegetation must be self-sustaining, and if vegetation fails, the revegetation process must be repeated.

Landfills and Waste Disposal Sites

Two major types of waste-disposal sites are used. One is the typical solid waste landfill used to bury municipal waste. These landfills have become large, high, and often involve a large flat top area with steep sideslopes. A major requirement is that surface runoff from the top is collected and safely conveyed to the bottom in well-armored channels (Figure 7.4). Diversions are used to prevent surface runoff flowing from an upper area onto the steep sideslopes, where very serious rill and gully erosion can occur. With slope lengths originating at the top of the sideslopes, erosion can be excessive. Diversions or terraces can be placed on the sides of the landfill, but care must be used. If the grade of the diversion is too flat, deposition will occur, restricting the diversion's transport capacity, causing overtopping, and gully erosion that exposes waste material. If the diversion grade is too steep, erosion can occur in the diversion itself, resulting in gully formation.

Establishing permanent vegetation is required for long-term erosion

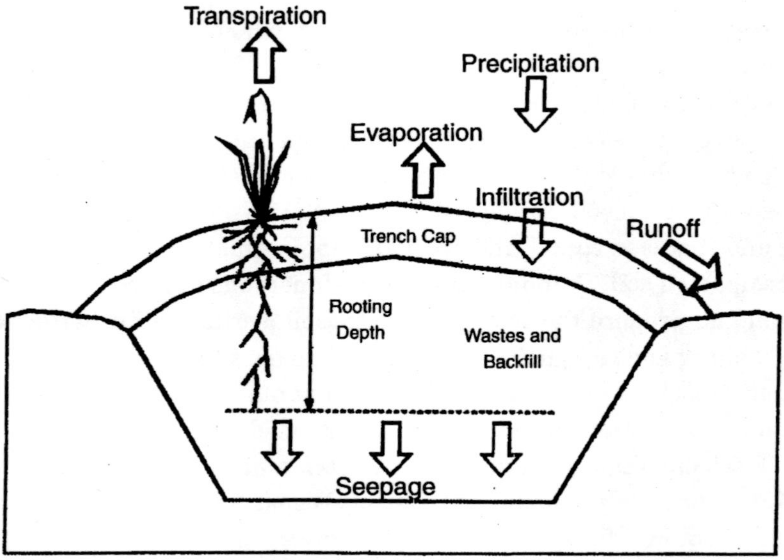

Figure 7.5 Design considerations for a waste disposal site. (From Nyhan and Barnes, 1989.)

control on landfills. However, typically poor soils make establishing and maintaining adequate vegetation very difficult. The vegetation must be chosen with consideration of the soil and local climatic conditions. However, temporary vegetation can be used effectively to assist in the establishment of permanent vegetation. Temporary mulch or rolled manufactured material might be used to control erosion in especially troublesome areas. Routine inspections are required and routine maintenance, such as mowing, is expected.

Figure 7.5 illustrates the hydrologic and erosion-control design considerations for a trench cap used at a low-level radioactive disposal site. Because the site must maintain the integrity of long-duration disposal, and because the sideslope must not erode to expose buried materials, the site must be designed very carefully. Furthermore, seepage from the base of the burial site must be prevented. Thus, the plants on the cap must be selected and established to provide an equilibrium between all phases of the hydrologic balance and to protect the site from erosion, which might expose the buried materials (Nyhan and Barnes, 1989). In some climatic situations, the trench bottom is lined with an impervious material (often, clay layers sandwiched between gravel drain materials) to eliminate groundwater pollution problems. The important disposal site criteria are to ensure stability on the cap sides and to ensure the disposal of water from the cap top (Figure 7.5). Thus, in addition to the erosion-control criteria, hydrologic considerations must be included at the disposal sites.

Summary

Table 7.3 summarizes water erosion-control practices, but many of the practices have a similar effect on wind erosion. The effect of these practices on wind erosion typically is to reduce erosivity more than reducing erodibility. Erosion is determined by two factors: (1) erosivity of the rainfall, runoff, and wind; and (2) erodibility of the soil. Erosivity is a measure of the forces applied to the soil that cause erosion, and erodibility is a measure of the susceptibility of the soil to erosive forces. It is the inverse of the soil's ability to resist erosion. Erosion control can be achieved in only two ways: (1) reduce erosivity (reduce the forces applied to the soil), or (2) reduce erodibility (reduce the susceptibility of the soil to erosion and increase the capability of the soil to resist the forces applied by the erosive agents).

The values in Table 7.3 compare soil loss under specific practices to soil loss from a continuous-tilled bare fallow unit-plot condition, as discussed in Chapter 5. The unit-plot condition is near the maximum erodible condition. A value for erosion-control effectiveness of 0.1 means that soil loss following application of an erosion control practice is 10% of the soil loss from the unit plot. An erosivity value of 0.5 means that the erosivity applied to the soil with the erosion control practice is one-half that applied to the soil with the unit-plot condition.

The values in Table 7.3 are very approximate. They vary with many factors, such as (1) location (e.g., climate), (2) slope steepness, and (3) biomass production level. These values are provided to indicate how erosion control practices vary in their effectiveness and how they affect erosivity and erodibility in a general way. These values should not be used in conservation planning. Erosion-prediction technology that takes into account simultaneously the many factors that affect erosion should be used for conservation planning.

Many combinations of conditions exist. The assumption for the effect of traffic is that a well-established grass cover existed before the traffic began. The values in Table 7.3 for traffic depend on the conditions that existed before the traffic started. If the soil was bare to begin with, traffic has far less impact than if an excellent grass cover was present.

Table 7.3 Cultural Management for Interrill and Rill Erosion Control

Land Use	*Practice*	*Erosion-Control Effectiveness*	*Erosivity Factor*	*Erodibility Factor*	*Comments*
Cropland	Clean tilled row crop/small grain	0.2	0.6	0.3	Typical farming practice without conservation
	Dense vegetation (hay, pasture)	0.005–0.03	0.1–0.4	0.05–0.08	Low-growing, dense canopy; produces some litter
	Crop rotation (row crop/small grain with hay)	0.03–0.2	0.4–0.6	0.08–0.3	Combination of row crop/small grain with hay crop
	Reduced till	0.1	0.4	0.25	Form of conservation tillage; about 35% residue cover at planting
	No-till	0.02	0.2	0.1	Form of conservation tillage; no tillage except immediately in row; about 60% residue cover at planting
	Manure application	0.15	0.55	0.25	Heavy application of manure before spring tillage
	Cover crop	0.30	0.8	0.4	Second crop planted after low-residue crop; provides additional cover and biomass
	Without cover crop	0.45	0.9	0.5	

Grazing lands	No grazing	0.001–0.05	0.02–0.35	0.05–0.15	Natural, wildland condition
	Light grazing	0.005–0.10	0.1–0.5	0.04–0.2	Grazing slightly reduces canopy and litter cover
	Heavy grazing	0.1–0.4	0.5–1.0	0.2–0.4	Grazing so intense that much of vegetation is killed
	Hay removal	0.01–0.03	0.16–0.3	0.06–0.1	Much of aboveground biomass removed during summer for hay; regrows during winter
	Animal lounging areas are typical	0.7	1.0	0.7	Areas where animals congregate for feed or shade
Disturbed forestland	Bare cut	0.45	1.0	0.45	Road cut
	Bare fill, smooth	1.0	1.0	1.0	Road fill
	Lightly disturbed	0.01	0.16	0.05	Typical of areas where only traffic has been to machine-cut individual trees
	Moderate disturbance	0.1	0.5	0.2	Repeated traffic area
	Heavy disturbance	0.5	1.0	0.5	Intense traffic, as in landings and skid trails
	Light site preparation	0.1	0.3	0.3	Light burial of debris by tillage
	Intensive site preparation	0.35	0.7	0.5	Removal of almost all debris by raking
	Recovery period	0.05	0.35	0.15	Weed/grass invasion
Construction sites/ reclaimed land/ landfills	Fresh cut	0.45	1.0	0.45	Immediately after operation
	Fresh fill, smooth	1.0	1.0	1.0	Immediately after operation

Table 7.3 *(Continued)*

Land Use	*Practice*	*Erosion-Control Effectiveness*	*Erosivity Factor*	*Erodibility Factor*	*Comments*
	Fresh fill, rough	0.8	0.9	0.9	Immediately after operation
	Mulch	0.15	0.11	0.9	Heavy cover of straw mulch, anchored to soil
	Temporary seeding with mulch	0.06	0.4	0.15	Temporary vegetation, small grain
Construction sites/reclaimed land/landfills	Permanent moderate cover	0.02	0.2	0.1	Established cover but of moderate quality and production because of poor soil conditions
	Permanent grass	0.005	0.1	0.05	Excellent grass
Military training	Light traffic	0.02	0.2	0.1	Light soil and vegetation disturbance
	Moderate traffic	0.3	0.85	0.35	Much of vegetation has been killed; much exposed soil
	Heavy traffic, smooth	1.2	1.0	1.2	Almost all vegetation killed; almost totally bare ground
	Heavy traffic, rough	0.7	0.85	0.85	All vegetation killed; no cover; very rough
Parks/recreational areas	Moderate foot traffic	0.15	0.6	0.25	Some vegetation killed and some soil exposed

Heavy foot traffic	0.3	0.85	0.35	Much of vegetation killed; moderate soil exposed
Light vehicular traffic	0.02	0.2	0.1	Much of vegetation killed; moderate soil exposed
Moderate vehicular traffic	0.3	0.85	0.35	Most of vegetation killed; much soil exposed
Heavy vehicular traffic	1.2	1.0	1.2	All vegetation killed; all soil exposed; loose soil layer (dust)
Mulch	0.15	0.11	0.9	Heavy, anchor mulch applied to reseed disturbed areas
Temporary vegetation	0.15	0.6	0.25	Quick-growing vegetation to provide immediate erosion control while permanent vegetation becomes established
Permanent vegetation	0.005	0.1	0.05	Vegetation approach for soil, climate, use, and maintenance

CONTROL OF CONCENTRATED-FLOW EROSION

Almost all areas where interrill and rill erosion are concerns include concentrated-flow areas (Chapter 3) that collect runoff in a defined channel network and deliver it to a watershed outlet. Farm fields, construction sites, and reclaimed lands commonly include several of these small watersheds. The boundaries of these sites may not match the watershed boundaries and thus require special consideration.

Erosion by concentrated flow is related directly to the shear stress applied to the channel boundary by the flow and inversely related to the resistance of the soil at the channel boundary. The shear stress applied to the soil is related directly to water discharge rate and channel grade and inversely related to hydraulic roughness of the channel boundary. The soil resistance is greater when the soil has not been mechanically disturbed than following disturbance.

Reduction of Discharge Rate

The discharge rate in concentrated-flow channels within field-sized areas is a function of the overland-flow runoff rate entering the channels. A change in land use can change the runoff rate significantly. A frequent problem in urban development is that the development creates a major increase in impervious areas. This dramatically increases runoff rate in channels that drain these areas. Such increases in discharge frequently cause channels that were stable before development to erode. Water-retention basins on the site can be used to decrease the discharge rate. These basins can serve as sediment-control areas during construction. After the development is completed, the main purpose of the basins is to retain runoff, which reduces the downstream discharge rate and channel erosion.

The same principle is used on farm fields where concentrated-flow (ephemeral) gully erosion is a problem. A system of structures known as *tile outlet terraces* are installed (USDA, SCS 1978). The terraces intercept surface runoff and move the water and sediment to control basins located in the concentrated-flow area. Water from the basins is released to an underground tile line at a controlled rate. In this system, the maximum discharge rate in the concentrated-flow channel is decreased in relation to the spacing of the structures along the ephemeral gully. In addition to controlling ephemeral gully erosion, these structures also trap sediment.

As much as 95% of the sediment reaching these small basins can be trapped (Laflen et al., 1980).

Reduction of Grade

The erosivity of channel flow can be decreased by reducing the channel grade. Terrace channels are constructed on a flat grade (about 0.5%) to cause deposition. The transport capacity in the terrace channel is less than the sediment load coming into the channel. As the channel grade is increased, deposition decreases as transport capacity increases. At steeper grades, the shear stress of the flow exceeds the soil's critical shear stress and the flow begins to erode the channel. Diversions are designed to be on a sufficiently steep grade that deposition does not occur, but not so steep that erosion occurs (Haan et al., 1994).

Concentrated-flow areas within the natural landscape may be on such a steep grade that erosion occurs in them. A drainage-system design may not be possible for channels with a sufficiently flat grade to prevent erosion. In certain land-use scenarios, channel grade can be reduced by installing a series of drop structures along the channel (Little and Murphey, 1982; Shields et al., 1995). Drop structures must be designed and constructed carefully to avoid excessive erosion in the plunge pool. An area where water velocity is dissipated in a ponded area, on the downstream side of a drop structure, provides some energy control. This area can be protected with rock or other nonerodible material to prevent excessive erosion. The control edge of the drop structure and the approach must also be constructed so that the control edge is stable and the banks guide the flow into the structure. The structure must be sufficiently large that runoff does not move around the structure, or if flow does move around the structure, a channel is available to transport this flow without erosion destroying the structure (Figure 7.4).

Concrete and rock chutes are used as drop structures. Elaborate chutes include energy dissipators at the end of the chutes. These energy dissipators eliminate erosion at the outlet of the structures, and they also allow using shorter structures, which reduces costs. Logs, sheet piling, corrugated metal, and other inexpensive materials also can be used to construct grade control and drop structures (Figure 7.4), provided that stability is ensured. The structure must extend beneath the channel perimeter sufficiently far to be stable after the flow erodes a plunge pool. The structure must also withstand the upper-side soil pressure.

Control of Headcut Erosion

Gully erosion frequently occurs as headcuts that move upstream in a concentrated-flow area (Chapter 3). A difference in base level where one channel enters another channel may create an overfall. Another example is where runoff from a field enters a drainage ditch. Flow at the overfall is very erosive, which can initiate a headcut. Drop structures such as those used to reduce the energy grade along a channel can be used to protect an overfall. Outlets from these structures need to be protected from downstream scour.

Another way to control headcuts in a gully involves a gully plug, which is basically a dam placed across the gully just below the headcut (USDA, NRCS, 1997c). In many such instances, poor construction causes flow to go around the structure or undercut the bank. The flow erosivity over the headcut is dissipated by the ponded water from the dam that inundates the headcut. The ponded area fills with sediment over time. The outflow device in the dam becomes, in effect, a drop structure.

In still other instances, runoff in a field can be transported to a permanent channel through a pipe installed to drop water from the field to the permanent channel (Figures 7.2 and 7.4). The inlet to the pipe must be designed to ensure a smooth transition from the field to the pipe without scour.

Protecting the Channel Boundary with Channel Linings

Although tile outlet terraces are effective runoff management and disposal systems, they are also expensive and can interfere with farming operations. Sometimes conversion of land from clean tillage to conservation tillage is sufficient to eliminate or greatly reduce ephemeral gully erosion. An alternative for controlling ephemeral gully erosion is to create a grassed waterway (Ree, 1949). Also, grassed waterways typically are used as an outlet channel to collect flow from gradient terraces and to carry the runoff downslope (Figure 7.2). Serious gully erosion often develops where terrace channels simply discharge on the side of a hillslope. Other locations where a grassed waterway can be used is for road drainage, airport drainage, and urban developments.

Temporary channel linings are used on construction sites for erosion control during construction and to protect the channel during establishment of permanent grass. These linings typically are manufactured roll products or blanket materials (Figure 7.2). The materials are designed to decompose during about a two-year period. The channel surfaces must be prepared carefully so that the materials make good contact with the soil everywhere along the channel; otherwise, water will flow between the liner and soil, causing erosion and failure. Thus, the lining material must be anchored into the underlying soil.

Channels also can be lined with rock riprap (Figure 7.2). The rock must be graded in sizes. Large rock provides stability because the weight is sufficiently large to resist movement by the flow. The small rock sizes fit into the openings between the large stones. Without these finer materials, the flow between the stones would erode the soil, causing a failure. In addition, a graded gravel or filter fabric is placed between the rock and the soil. Concrete and asphalt can also be used to line channels, but as with rock, these materials are expensive, require maintenance, and often are considered unattractive. Generally, grassed-lined channels are preferred, but a disadvantage of grass is that it must be maintained.

Stream Bank Protection

Sometimes stream banks become unstable because of rapid lateral erosion, especially on the outside of meander bends (Bowie, 1995: Shields et al., 1995). In some cases, the banks retreat because of erosion at the base of the channel wall, resulting in sidewall collapse (Chapter 4). One approach is to grade the channel bank to a steepness such as 2 (horizontal) : 1 (vertical) and establish a grass that can resist the erosive forces of the flow. In more difficult cases, willow trees, which can tolerate high velocities along the stream banks, are planted (Bowie, 1995; Shields et al., 1995). Grass also is interseeded between the trees. The vegetation must be appropriate for the local climate and soil conditions. In the most severe cases, rock riprap can be placed on the banks, but rock is expensive and may not be aesthetically pleasing. Also, rock-filled gabions (rock filled wire baskets) can be used (USDA, SCS, 1978).

SEDIMENT CONTROL

Sediment loss from a site can be controlled in one of two ways. The first way is to control erosion, and the second way is to reduce transport capacity at the boundary of the site, which will lead to deposition of the sediment load. The preferred approach is to control erosion. Controlling erosion protects the landscape and soil from degradation. As noted previously, coarse sediment leaving the site is preferred because this sediment can be easily deposited farther downstream, and coarse sediment degrades water quality less per unit mass than does fine sediment. Fine sediment has more surface area to adsorb chemicals. Controlling sediment loss by controlling detachment generally leaves coarser sediment than where sediment loss is controlled by a reduced transport capacity. However, erosion control cannot always be used to reduce sediment delivery, such as on construction sites. Construction frequently exposes the land to erosion for at least a few months. Some form of sediment control usually is required by law. Also, sediment-control requirements may be more restrictive than erosion-control requirements on certain farmland. In that situation, a combination of practices will be used; one for erosion control to protect the soil against degradation by erosion and another for sediment control. If runoff leaves the site by way of concentrated-flow areas, the sediment control measures can be placed at the outlet points. If some of the runoff leaves the site as overland flow, control measures must be placed along the site boundary.

Sediment Control for Overland Flow

On construction sites, reclaimed mined land, and other lands where the topography can be shaped, reducing overland-flow transport capacity and causing deposition is possible by shaping the lower portion of a hillslope (Chapter 4). A concave shape where the lower end of the slope is very flat (1% steepness or less) causes significant deposition. Space is a problem on many construction sites so that a long concave hillslope cannot be constructed within the available property. However, just a 25-ft (8-m)-long 1% slope segment at the end of a 175-ft (50-m) hillslope that is 17% in steepness will cause 80% of the sediment load to be deposited. Such a slope segment is relatively short. The problem with the short hillslope segment is that deposition can steepen the last segment, increasing the runoff transport capacity, decreasing deposition and increasing sediment loss. Although the concave hillslope segment may not be sufficient by itself, it can be used in conjunction with a filter strip and other sediment-control measures.

A filter fabric commonly known as *silt fences* often is used to reduce the transport capacity of runoff and cause deposition. These barriers can be highly effective, but the fences are often not installed and maintained properly. The ponded water and sediment behind the fence puts great stress on the fence and its supports. If the supports are not strong and the fabric well attached to the supports, the fence fails. Fences also commonly fail when runoff flows under them. This failure is avoided by burying a portion of the fabric.

Another barrier type (Dabney et al., 1993) is stiff-grass hedges planted in narrow rows about 18 inches wide. These stiff grasses can trap up to about 90% of the sediment. The problem with stiff grass is rapidly establishing a uniform stand. Use of fabric fences often is easier than installing grass hedges. If vegetation is used as a barrier, stiff grasses such as vetiver grass, switch grass, and cane beardgrass are more effective than common grass such as bermuda and bluegrass. The stiff grass stands up to the flow to maintain its effectiveness. Also, the stiff grass can recover after being inundated by deposition.

Grass strips more than 15 ft (5 m) wide are used on cropland to control sediment yield and to improve water quality. These strips do not need to be very wide to control sediment because most of the deposition is in the backwater on the upper side of the strip (Chapter 4). Over time, the backwater area fills with sediment and sediment is transported into the grass. Thus, the grass strip needs to be sufficiently wide to avoid inundation with sediment. The grass must be sufficiently vigorous to recover after deposition. Typically, the width of the strip is determined by the width required by water quality purposes rather than for sediment control.

Gravel bags and hay or straw bales are sometimes used as flow barriers. Straw bales are susceptible to runoff flowing between the ends of the bales where they are butted together. Bales must be firmly staked in place but can still unravel. Gravel bags work better than straw bales, and if placed sufficiently high and have a wide base, they can be more stable than a fabric fence. However, the fabric fence is much easier and cheaper to install. The effectiveness of these practices depends on the quality of the installation and maintenance (USDA, SCS, 1977).

One of the major problems with barriers is that site boundaries are seldom on the contour. If the barriers are even on a slight grade, water can flow along the barrier without much deposition. Even without deposition, the barriers still serve the useful function of keeping runoff within the site boundaries and directing the runoff to a collection point.

Terraces with a low grade can be used to control sediment by causing deposition in the terrace channel. The terraces must be sufficiently high to retain the deposited sediment and still provide sufficient flow capacity.

On cropland, routine terrace maintenance is required to remove the deposited sediment from the channel area to maintain the flow capacity (Foster and Highfill, 1983).

Sediment Control in Concentrated-Flow Areas

Straw bales and gravel bags are sometimes placed in concentrated flow areas to control erosion (USDA, SCS, 1977). However, this practice often is relatively ineffective. Even when the bales stay in place, flow can go around the ends of the bales, causing scour. In highway construction, specially formed bales are often used. The bales are long, round, flexible, and conform to the topography. These bales must be anchored in the channel to prevent them from rolling downslope.

The most effective sediment control device is a sediment-control basin. When these basins are properly installed, they can trap up to 95% of the sediment entering them (Laflen et al., 1980). Also, sediment-control basins can be put in series, but the second basin is much less effective in trapping sediment than the first basin because the first removes most of the coarse sediment and the fine sediment leaving the first basin is not easily deposited in the second basin (USDA, NRCS 1997c). Although sediment basins are highly effective at removing sediment from the runoff, the remaining sediment load is enriched in fine particles. Finally, sediment-control basins are expensive and require maintenance.

The sediment trapped in basins that are a part of terrace systems on cropland can be removed with farm equipment as a part of routine farming operations. In general, the sediment-bench accumulation is considered to be beneficial. However, an important design consideration of sediment basins for construction sites involves determining their size requirement (USDA, NRCS, 1997c). If the basins are too small, they fill quickly and rapidly lose their trapping efficiency. If the basins are built too large, the cost is excessive. Given that the basin may be in service only for a short time on a typical construction site, the contractor may be willing to take a reasonable risk in sizing the basin. If the basin fills with sediment, the contractor can remove and spread the sediment. However, a suitable place must be available to spread the sediment. Sometimes sediment basins are cleaned when the development is complete and the basin is used as a water-retention basin to control downstream discharge rates.

Watershed Sediment Yield

Erosion control on a watershed scale is very complex. Thus, erosion scientists and engineers frequently resort to using a relationship expressing the amount of sediment passing a point on a stream to the amount of sediment removed from an upstream area for conservation and reclamation planning and design. The ratio of these two values is referred to as a *sediment delivery ratio*.

Many estimates of sediment-delivery ratios have evolved from past measurements. As a general rule, sediment delivery ratios have been observed to decrease with increasing watershed area. When erosion-control practices are present in a watershed, the decrease in delivery ratio with watershed size often is difficult to determine. When a reservoir is used for water supply or flood control on large watersheds, there is a need to design the reservoir to accommodate sediment storage (Renfro, 1975; Roehl, 1965). In such instances, sediment-delivery ratios can be used with caution in the absence of watershed data.

WIND-EROSION CONTROL

Wind-erosion control principles are essentially the same as those for water erosion. Many of the practices used to control water erosion also control wind erosion, with adjustments for the differences between wind-erosion mechanics and water-erosion mechanics (Chapter 4). For example, roughness has a greater effect on wind erosion than on water erosion. For water erosion, the preferred canopy is one that is short, whereas a tall canopy is preferred for wind-erosion control. Wind erosion damages the soil by reducing soil depth and by selectively removing clay and silt particles while leaving sand particles. Controlling average annual wind erosion is usually sufficient to protect against this damage to the soil.

Other on-site damage caused by wind erosion is the abrasion of seedlings by windblown sediment. In severe cases, a crop must be replanted, while in less severe cases, yield is significantly reduced: In either case, considerable economic loss may occur. The important variable is abrasion during the critical periods before the seedlings become sufficiently developed to resist the effects of the windblown sediment. Erosion control is essential during this seedling period. Controlling average annual wind erosion may not protect seedlings.

Figure 7.6 Wind erosion reduces visibility and causes deposition. (Courtesy of USDA, NRCS.)

Sediment leaving a wind-erosion source area can cause downwind damages ranging from sediment deposition to respiratory problems. Sediment can be deposited in fence rows and road ditches immediately adjacent to the source area, as illustrated in Chapter 3. Windblown sediment can reduce visibility for long distances away from the source area. Also, dust in windblown sediment can cause medical problems, be a general nuisance, and be a carrier of polluting chemicals (Figure 7.6). Trapping coarse sediment at the site boundary is possible but not especially effective or feasible in the long term. The best way of controlling the windblown sediment loads, including coarse as well as fine particles, is to control erosion at its source and prevent sediment movement (Fryrear and Skidmore, 1985).

Wind Erosion Control Principles

Reducing the erosivity of the wind has a much greater relative effect on reducing wind erosion than does reduction of the erosivity of rainfall and runoff. Therefore, erosion-control practices that reduce erosivity of the wind at the soil surface generally are preferred over the practices that reduce soil erodibility. One of the major benefits of preventing detachment in the first place is that wind erosion is deprived of a major erosive agent in that windblown sediment abrades the soil, producing additional sediment that itself becomes an erosive agent.

The major cultural practices to control wind erosion are those that provide a canopy cover, ground cover, and leave the soil rough. Biomass is more effective at controlling wind erosion when it is standing than when it is flat on the soil surface (Bilbro and Fryrear, 1994). But even short vegetation reduces wind velocity significantly at the surface. A high density of roots near the soil surface also can decrease soil erodibility and be effective in reducing wind erosion. Maintaining a vegetative cover, canopy cover, and ground cover is the key erosion-control practice. Crop rotations that combine erodible crops with nonerodible crops effectively control average annual erosion. However, considerable risk is involved during the years of the erodible crops that wind erosion will cause plant abrasion or problems with off-site sediment delivery. Conservation tillage systems that leave large amounts of plant residue on the soil surface and leave the surface rough are effective wind-erosion control practices. The peak wind erosivity varies during the year, and the objective is to have good vegetative cover at the time of peak erosivity. Another temporal variable is soil moisture. A dry surface soil is much more susceptible to wind erosion than is a moist surface. Therefore, an erosion control objective is to have good vegetative cover on the land during dry periods, especially if wind erosivity is high during those periods.

When placed perpendicular to the prevailing wind, ridges can be highly effective at reducing wind erosion. The other major practice involves barriers, known as *wind breaks,* around the site boundaries. These barriers can be strips of tall stiff grass, agronomic crops such as corn, and permanent trees. Some grasses, such as big sacaton and even some sorghums, provide wind-erosion control. These barriers control erosion by reducing the erosivity of the wind for some distance into the site. Barriers can also be used to cause deposition. A snow fence used to deposit snow away from a roadway or a house is based on the same principles of wind-erosion control. These barriers can be effective at removing the coarse particles but have little effect on the suspended load. Barriers that are tall and dense are most effective at reducing wind erosion.

Tall vegetation has more effect in reducing wind erosion than short vegetation of the same density. However, replacing shrubs with short grass substantially reduces wind erosion because the short grass usually is much more dense than taller shrubs. In shrub-dominated lands, the intershrub spaces afford areas for wind erosion.

Wind-Erosion Control on Construction Sites

Wind erosion on construction sites frequently is a problem. The problem is accentuated because of population proximity to such sites. The wind-blown sediment causes respiratory problems and may be deposited near high-value property. Control of such erosion may involve use of irrigation, establishing temporary vegetation, or using a soil stabilizers described by Armbrust and Lyles (1975).

Wind-Erosion Control on Abandoned Irrigated Fields

In the dry climate of the southwestern United States, wind erosion from abandoned irrigated cropland causes major problems. Cessation of irrigation and abandoning farming on these lands requires special wind-erosion control. Wind erosion on these areas leads to severe automobiles accidents (e.g., in Arizona, California, and New Mexico) as drivers slow. Laws now require a "native" vegetation cover when irrigation terminates. The last irrigation application or two are used to germinate and establish a vegetation to control wind erosion. State highway departments also have established elaborate warning systems alerting travelers of reduced visibility.

Wind-Erosion Control on Rangelands

Plant communities such as mesquite, creosotebush, and tarbush are the vegetation types prevalent on deteriorated rangelands in the southwestern United States. Coarse-textured windblown sediment is deposited beneath the plant canopy (Hennessy et al., 1986). The area and height of dunes have been measured to increase significantly. Within a dune, sand deposition dominates the vegetation canopy, whereas the finer-textured particles are moved from the eroding sites (Gibbens et al., 1983). Erosion is controlled on these lands by killing undesirable plant communities with a herbicide and replacing them with desirable grass species. Sand movement was reduced 95% with the treatment. Production of desirable plant communities provided wildlife habitat and forage for cattle production while reducing wind erosion (Herbel and Gould, 1995).

SUMMARY

Land-disturbing activities expose the soil to raindrop impact, surface run-off, and wind. Excessive erosion can occur that degrades the soil and produces the sediment that causes off-site damage. Both erosion and sediment control are required. The preference is to control both erosion and sediment by controlling erosion, especially for wind erosion. However, situations exist where erosion control may not be possible, so that sediment-control practices such as barriers and sediment-control basins are required. Sediment control may not provide adequate erosion control to protect the soil and the land. Water and wind erosion control are accomplished by adequate canopy and ground cover during the most erosive times of the year. Land-disturbing activities should be restricted to non-erosive periods when possible, or temporary cover provided. Agronomic practices for erosion control such as contouring or the new conservation practices in modern farming offer great promise at minimal costs.

SUGGESTED READINGS

Agnew, B. 2000. *Practical Approaches for Effective Erosion and Sediment Control*. International Erosion Control Association, Steamboat Springs, CO.

El-Swaify, S. A., E. W. Dangler, and C. L. Armstrong. 1982. *Soil Erosion by Water in the Tropics*. College of Tropical Agriculture and Human Resources, University of Hawaii, Honolulu, HI.

Troeh, F. R., J. A. Hobbs, and R. L. Donahue. 1991. *Soil and Water Conservation*, 2nd. ed. Prentice Hall, Upper Saddle River, NJ.

U.S. Department of Agriculture, Natural Resources Conservation Service. 1997. *Ponds: Planning, Design, Construction*. USDA Agricultural Handbook 590. U.S. Government Printing Office, Washington, DC.

U.S. Department of Agriculture, Soil Conservation Service. 1977. *Erosion and Sediment Control in Developing Areas: Planning Guidelines and Design Aids*. USDA Printing, Columbia, SC.

8

Land Conservation

The land and its soil are used to produce food, fiber, timber, coal, and minerals. Land provides space for housing, transportation, industry, commerce, recreation, and waste disposal. Land supports diverse wildlife habitats and often provides attractive aesthetic qualities. Land price is one measure of land value, but the value of wildlife habitat and aesthetic qualities is not easily determined by economic measures. Some lands are so special that they are protected and maintained as parks and wilderness areas in pristine condition to protect rainforests, old growth timber, endangered species, and other unique resources.

Land ownership is shared between the public sector (government) and the private sector. With the exception of protected public lands, much public land is used for mining, timber harvest, and grazing. The use of public lands frequently is debated. Major issues include payment levels by private users for the use of public lands (e.g., timber harvest and grazing fees) and the management of these lands (e.g., timber harvest method and grazing intensity). Users of public lands prefer few restrictions, but others place a greater value on land preservation than on utilization.

Owners of private lands generally are free to use their lands as they choose as long as they do not harm others, such as off-site sediment delivery that degrades air or water quality. Public decisions to protect and restrict the use of private lands must consider the rights of private ownership versus the public good. For example, to what degree should privately owned wetlands be protected for the good of society in contrast to the landowners' right to develop them for private gain?

Land is widely recognized as a global resource, as well as national resources, that requires long-term protection for society's well-being. However, differing values among competing interests for land use often are the deciding factors in land conservation rather than scientific and technical factors. Land conservation involves consideration of cultural, social, economic, environmental, and political factors.

Wealth and natural resources are not uniformly distributed within countries and between countries. Intensive farming to meet food demands in a country with great population pressures can excessively degrade land, while agriculture in other parts of the world produces abundantly without land degradation. A good land-conservation strategy would be to retire the severely degraded land and offset lost crop production by transfer of excess agricultural products from the surplus areas to the deficit areas. However, this conservation strategy is contingent on the country with the highly degraded land having sufficient wealth to purchase, transport, store, process, and distribute food and fiber products and on the willingness and ability of other countries to share these products and costs. Similarly, a poor economy within a country prevents public investment in soil conservation programs and individuals from investing in soil conservation on their land. In addition, farming methods often are passed down to succeeding generations. Changing to conservation farming methods may be contrary to custom, tradition, and peer pressure. In some cultural and political systems, land is divided into smaller units as it passes from generation to generation, which can make land conservation difficult when multiple tracts of land are involved (Hudson, 1981b). Many countries do not have an infrastructure of research, development, education, and technical assistance to support land conservation. The need for conservation often does not match the opportunity for conservation.

Although land is damaged in numerous ways, our particular interest is degradation by soil erosion. Any land is potentially threatened where accelerated erosion occurs. Land must be protected against excessive soil erosion to remain a viable resource indefinitely. A few years of careless use can greatly accelerate soil erosion that degrades the land permanently, especially for food production (Figure 8.1). Some lands degraded by erosion can be restored, but at a prohibitive cost, including removing the land temporarily from food production. Countries with severe population pressure may not have the luxury of permanently, or even temporarily, retiring land from food production or may not have the resources to apply soil conservation to the land if the land stays in production.

Figure 8.1 Severe erosion of agricultural lands reduces future productivity. (Courtesy of USDA, NRCS.)

PUBLIC CONSERVATION PROGRAMS

Every citizen of Earth depends on the land and its products, and almost everyone is directly or indirectly affected by environmental quality through health, economics, or recreation. Maintenance of environmental quality, including land, soil, air, and water quality, is a principle and value of modern society. Soil conservation protects the land from excessive erosion while utilizing land to meet the needs of modern society. Thus each person has an obligation to land conservation.

Public policy plays a major role in soil conservation for both public and private lands. Although the principal responsibility for soil conservation lies with those who use the land, the public has assumed a joint responsibility with both landowners and users by sharing the costs of conservation. The public, in turn, requires that certain standards be met in exchange for public conservation funding. In simple terms, tax money collected from the general citizenry is provided for soil conservation through government programs in a joint public–private partnership. This partnership evolved in the 1930s in the United States and continues to the present (Napier, 1990).

This public–private partnership for private-land conservation has several features:

1. *Land stewardship*. Those who own and use land are expected to exercise stewardship to improve the land during their tenure and to deliver it to succeeding generations in better condition than they received it. This conservation ethic is encouraged and promoted by both public and private organizations.
2. *Research and development*. Research is conducted by government agencies and academic institutions to understand the ways in which erosion occurs, erosion degrades soil, sediment is transported from erosional sources and causes downstream damages, and erosion and sediment delivery can be controlled. Practical erosion- and sediment-control practices are developed through research and refined by field application.
3. *Education*. Developing and proving soil-conservation technology is only a part of the effort; this technology must be implemented. Information must be delivered to a highly diverse and dispersed set of decision makers, who must receive the information and be convinced to implement soil-conservation measures. Public and private organizations conduct a wide array of educational and technology transfer programs. Articles published in newspapers and magazines, and public meetings are major sources of soil-conservation information. A well-proven education method for encouraging soil conservation is an on-site demonstration of soil-conservation practices. Farmers are greatly influenced by their peers and generally are cautious about whether a particular practice will work on their farms. Arrangements are made for a small group of farmers in a community to apply new conservation technology on their farms. Neighbors are invited to field days where the new technology and its success are displayed. Similar demonstrations have been conducted on forestlands and erosion-control products for construction lands are displayed routinely at trade shows.
4. *Technical assistance*. Most erosion- and sediment-control practices must be properly selected to fit site-specific conditions. Government agencies establish local offices in cooperation with local organizations to provide technical assistance to landowners and users for the design, installation, and maintenance of soil-

conservation practices. This assistance typically is provided as a part of established government programs where cost sharing is provided with the proviso that the soil conservation practices are installed and maintained according to certain specifications.

5. *Government programs*. Government conservation programs include both voluntary and regulatory programs. The voluntary programs typically provide cost-sharing funds as the incentive to participate. Linkage with unrelated but highly desired government programs that provide economic support is also used as an incentive to participate in soil-conservation programs. Participation in the soil-conservation program is mandatory as a condition for participation in economic support programs. Voluntary, incentive-based programs are used most frequently to promote soil conservation on private lands where the purpose is to protect the on-site soil resource, whereas regulatory programs often are used to control off-site sediment delivery. Regulatory programs typically require a permit to conduct land-disturbing activities and a mandated level of erosion and sediment control to maintain the permit and avoid fines. Some of these programs provide cost-sharing funds.

CONSERVATION PLANNING

Two forms of conservation planning are used to implement soil-conservation programs. One form is local planning to select erosion- and sediment-control practices for specific land units, and the other form is planning to develop and implement public policy. The objectives of conservation planning are to (1) identify lands that are eroding excessively, and (2) select practices to reduce erosion and sediment delivery to acceptable levels on the excessively eroding lands. Erosion- and sediment-control practices should be scientifically based; technologically feasible; allow the preferred land use; be profitable and cost-effective; fit the local cultural, economic, social, and political setting; and provide for long-term sustainability.

Geologic erosion rates are considered acceptable, but what erosion rate in excess of geologic erosion is excessive? One indication of excessive erosion is that the land cannot be used for its chosen purposes. For example, gullies on cropland interfere with farming and reduce cropland value. Excessive water and wind erosion degrade the productivity of shallow soils.

Rills and gullies interfere with revegetation and maintenance, such as mowing on highway embankments. Large amounts of sediment leaving an eroding site fill downstream water-conveyance channels and reservoirs. Fine sediment degrades water and air quality.

Before conservation planning can occur, a criterion for excessive erosion is chosen as the basis for selecting conservation practices. One criterion is a performance-based standard where the erosion- and sediment-control practices must reduce erosion and sediment-delivery rates to less than the standard under site-specific conditions. The other criterion is based on the application of *best management practices* (BMPs) where particular practices are chosen without full consideration of the resulting erosion and sediment-delivery rates.

Best Management Practice Approach to Planning

The BMP approach to conservation planning is to select erosion-control practices from a list of approved best management practices that are judged to provide satisfactory erosion and sediment control. The BMPs are chosen without full consideration of the resulting erosion and sediment-delivery rates and sometimes without tailoring the practices to site-specific conditions. The BMP conservation planning approach is favored in many regulatory programs because it is easy to administer and enforce (U.S. Environmental Protection Agency, 2000). Regulatory compliance requires only that the practices have been selected from an approved list and that the practices have been installed and maintained according to specifications. The BMP approach provides a relative degree of protection, such as a 75% reduction in soil loss rather than the reduction of soil loss to a specific quantitative standard.

Quantitative Standard to Protect On-Site Soil Resources

The most widely used quantitative standard for conservation planning to protect on-site soil resources from excessive erosion is the *soil-loss tolerance* concept described in earlier chapters. Soil-loss tolerance "denotes the maximum level of erosion that will permit a high level of crop productivity to be sustained economically and indefinitely" (Wischmeier and Smith, 1978). Conservation practices are selected to control average annual soil loss to a rate less than the soil-loss tolerance (T) value assigned to each specific soil. Because the objective is long-term soil maintenance, cumulative erosion over a long period, rather than erosion from a single or a few storms, is the basis for the soil-loss tolerance rate. High soil-loss rates are acceptable in any one year if the overall erosion rate is low.

Soils differ in their erodibility and the damage caused by erosion. For example, erosion of a shallow soil underlain by bedrock is far more serious than erosion of a deep loess soil. An eroded deep loess soil can be reclaimed by cropping-management practices, but when the soil over bedrock is lost by erosion, the damage is permanent, at least within human time frames. BMPs for erosion control may not be sufficient for such shallow soils. Instead, these lands should be maintained in a land use to keep erosion rates extremely low. In the short term, many soil-conservation practices are considered to be nonprofitable, so that the incentive is to apply no more erosion control than is necessary. The conservation-planning approach that emerged along with erosion-prediction technology in the 1940s was based on the quantitative soil loss tolerance rather than the relative BMP approach.

T values are assigned by the U.S. Department of Agriculture (USDA) according to specific soil conditions based on factors first formalized by USDA research scientists and field conservationists in a 1956 workshop. The *T* values, which have been assigned to almost all cropland soils and many other soils in the United States, range from 1 ton/acre (2 t/ha) per year for fragile soils to 5 tons/acre (11 t/ha) for soils not readily damaged by erosion. The factors considered in assigning a *T* value to a particular soil include (1) the rate of soil formation from parent material, (2) the rate of topsoil formation from subsoil, (3) reduction of crop yield by erosion, (4) changes in soil properties favorable for plant growth by erosion, (5) soil depth, (6) loss of plant nutrients and organic matter by erosion, (7) the likelihood of rill and gully formation, (8) sediment deposition problems within a field, (9) sediment delivery from the erosional site, and (10) the availability of feasible, economic, culturally and socially acceptable, as well as sustainable soil conservation practices (Mannering, 1981; McCormack and Young, 1981).

Although the soil-loss tolerance concept was originally developed for conservation planning on cropland, the concept also is applied to erosion control on landfills, reclaimed mine land, construction sites, and other highly disturbed lands, partly because no better standard has been developed, and because many of the principles embodied in the soil-loss tolerance concept apply to all lands. Although the soil-loss tolerance standard achieves many goals, reducing erosion to the soil-loss tolerance value may not sufficiently reduce sediment yield and dust emissions from erosional sites, will not protect against gully erosion where runoff is allowed to concentrate, and may not protect young seedlings from windblown sediment. Meeting the soil-loss tolerance standard is only one part of a comprehensive conservation plan.

Research data, scientific knowledge, and professional judgment are involved in setting soil-loss tolerance values. Those judgments in the 1950s were within the context of social and economic conditions and the agricultural technology available at the time. Soil-loss tolerance values periodically should be updated, but soil-loss tolerance values have become so institutionalized that suggestions of value change often are resisted (Esseks et al., 2000; Hall et al., 1985; McCormack and Young, 1981).

Relation of *T* Values to Soil Formation

Two soil-formation rates are considered in setting *T* values. The first is the rate of topsoil formation. A high-quality, well-aggregated surface soil layer greater than 10 in. (250 mm) that allows root growth, good aeration, and soil-moisture conditions, and contains an adequate supply of plant nutrients and organic matter, is required for vigorous plant growth. The rule of thumb is that about 30 years is required to form 1 in. (25 mm) of topsoil from subsoil with good cropping-management practices that incorporate significant amount of organic material into the soil. This formation rate equals about 5 tons/acre (11 metric tons/ha) per year, which sets the maximum allowable erosion rate (soil tolerance value) to maintain fertile and productive topsoil (McCormack and Young, 1981; Morgan, 1986). This rate of topsoil formation also seems to apply to well-vegetated reclaimed surface mined land (Schafer et al., 1979).

The other important soil formation rate is the formation of subsoil from parent material. The preferred *T* value is the erosion rate that just balances the rate of soil formation from parent material. This soil formation rate, however, is very difficult to determine because soil formation occurs within a soil profile and varies considerably with parent material and climate. An average soil formation from parent material seems to be about 0.5 ton/acre (1 metric ton/ha) per year, but rates can be as low as 0.05 ton/acre (0.1 metric ton/ha) per year and as high as 1.5 tons/acre (3 metric tons/ha) per year, depending on climate and other factors (Morgan, 1991). Soil-formation rates are considered to be an order of magnitude (one-tenth of) less than typical soil loss tolerance values, and therefore some scientists consider erosion at soil-loss tolerance values to allow a slow deterioration of the soil instead of the indefinite protection implied in the definition of soil-loss tolerance (Johnson, 1987; Lal, 1985, Troeh et al., 1991). However, controlling erosion rates to soil formation rates is not feasible with most current farming practices. Soil conservation policy-makers consider erosion rates higher than soil formation rates acceptable relative to present societal interests.

Relation of *T* Values to Crop Yield

Overall rooting depth is an important factor in plant growth because crop yield is determined primarily by soil-moisture storage in the rooting zone (Hall et al., 1985). Soil moisture for plant growth is reduced as erosion reduces soil depth above a dense horizon, such as a fragipan, parent material, or bedrock. Also, erosion removes plant nutrients and organic matter required for optimum crop yield. Thus, crop yield loss is a measure of erosion impact, and numerous field and analytical studies have shown that erosion reduces yield on certain soils (Hall et al., 1985; Schertz et al., 1985; Weesies et al., 1994). Adding fertilizers, manure, and other soil amendments offsets some plant nutrients and organic matter losses from the soil by erosion, but these additions increase farming costs and typically are assumed to be short-term rather than long-term, sustainable remedies.

The relationship of crop yield to erosion is complicated because new and improved plant varieties, fertilizers, and pest controls have greatly increased crop yield since the 1940s. Therefore, assignment of soil-loss tolerance values requires consideration of the advances in agricultural technology that offset the losses in crop yield caused by erosion, the costs of implementing new technologies, and whether agricultural systems based on these technologies are sustainable. Although technological advances do more than offset the productivity losses caused by erosion, the role of soil quality is frequently overlooked. Even with new technologies, the long-term productive advantage remains with those soils where soil quality is maintained and protected from degradation by soil erosion.

Relation of *T* Values to Plant Damage

Abrasion of plant seedlings by windblown sediment is one of the damages caused by wind erosion that can require expensive replanting, significantly reduce the yield, and degrade plant material quality. Controlling wind erosion to the soil-loss tolerance value does not necessarily provide adequate erosion control because soil-loss tolerance is based on long-term average annual soil loss, whereas this abrasion damage usually occurs during a particular season. A seasonal soil-loss tolerance could be established based on the extent to which a particular plant type is susceptible to abrasion damage. Erosion rate, characteristics of the sediment, wind speed, and susceptibility of particular plant species are factors considered in setting these soil-loss tolerance values.

Relation of *T* Values to Rill and Gully Erosion

If nothing else, rills and gullies are major nuisances. Whereas rills routinely can be obliterated by tillage on cropland, deep, incised gullies cannot be filled easily (Chapter 3). Although gullies can develop independent of rills, land management that controls rill erosion also reduces the likelihood of gully formation. The rule of thumb is that soil-loss tolerance values are set below about 7 tons/acre (15 metric tons/ha) per year, the interrill and rill erosion rate at which rills begin to form.

Relation of *T* Values to Conservation Technology and Other Factors

The soil-loss tolerance values set in the 1950s were based on the soil conservation technology available at the time that could be reasonably and profitably applied. By the mid-1980s, conservation tillage, based primarily on leaving residue from last year's crop on the soil surface, had been developed as a new erosion-control technology (Chapter 7). This technology largely replaced the widely used clean-till technology, and thus from one perspective, soil-loss tolerance values could be lowered because of the new conservation-tillage technology. However, when the 1985 Food Security Act was implemented, acceptable conservation systems (ACS) were identified that were based on double soil-loss tolerance values (2*T*) if ephemeral gully erosion was a problem in the field, or triple soil-loss tolerance values (3*T*) if ephemeral gully erosion was not a problem. Practices adopted from a list of ACSs (much like BMPs) were considered to be acceptable. A large number of farmers felt that achieving the soil-loss tolerance would cause too much economic hardship in the short run (Esseks et al., 2000). These farmers had sufficient political clout to influence policymakers, resulting in the use of higher soil-loss tolerance values (2*T* or 3*T*) and a BMP-type approach of selecting from a list of "acceptable conservation systems."

A major structural change occurred in U.S. agriculture that affected soil conservation. Crop rotations for erosion control include a densely vegetated crop, such as hay, that greatly reduces erosion (Chapter 7). The low erosion under the hay crop offsets the high erosion rate resulting from clean-tilled row crops such as corn, so that the overall average erosion rate is low. Such crop rotations were feasible in the 1950s because most farms produced livestock that could profitably utilize the hay. Today, because few U.S. farms have livestock, hay often is not profitable as a cash crop such as corn, soybeans, wheat, and cotton, except for certain specialized markets. Many farms only grow row crops, such as corn and soybeans, with no hay or pasture crops. Thus, little, if any, net reduction in erosion was realized under conservation tillage; it simply offset the increase in erosion rates that occurred when hay and pasture crops disappeared from many farms. This structural change in agriculture illustrates the complexity of assigning soil-loss tolerance values and the way these values, or at least how they are used, change through time.

Summary Statement for Soil-Loss Tolerance

The benefits of using soil-loss tolerance values as a conservation-planning tool are well recognized and accepted. Preferred soil-loss tolerance values will continued to be debated, and appropriately so. The soil-loss tolerance concept includes a significant socioeconomic element that cannot be reduced to hard science, and science cannot precisely determine soil-formation rates nor the physical and biological impact of erosion on soil. The value judgments associated with cultural, socioeconomic, political, and similar considerations vary among individuals. However, the feature that makes soil-loss tolerance such a powerful conservation-planning tool is that both technical and nontechnical elements are combined into a single number.

Standards for Sediment Control to Protect Off-Site Resources

Sediment control can be achieved in one of two ways: (1) erosion control that reduces sediment production, and (2) on-site sediment deposition induced by an impoundment, vegetative, or mechanical barrier. The sediment delivery rate (sediment mass delivered through time), concentration of sediment in water and air, and the physical characteristics of sediment are considered in setting criteria for allowable sediment delivery. The requirements of the resource affected determine the tolerance values. For example, filling of roadside ditches by sediment is an off-site impact. In this case, the sediment load delivered to the ditches by runoff from nearby areas should be less than the transport capacity of flow in the ditches, so that deposition in ditches is minimized. Similarly, deposited sediment fills downstream reservoirs. In addition to other factors, such as design flood magnitude, reservoir size is determined by the amount of sediment that will be trapped by the reservoir during its design life. Erosion control frequently is a part of the overall watershed-management plan when a reservoir is built.

Both the total amount of deposited sediment and the fine-size component in the sediment can degrade the ecological well-being of stream channels. The concentration of fine-size material is an important consideration in treatment of drinking water. Total maximum daily load and concentration of total suspended solids often are used to select erosion and sediment control practices to control the concentration of fine sediment in the water. The concentration of fines (dust) in the air produced by wind erosion may result in visibility and respiratory problems. The concentration of particles in air less than 10 μm (PM10) is a frequently used standard for the regulation of dust in air.

In contrast to protecting on-site soil resources where the time scale usually is the long-term average annual erosion, the time scale for protecting off-site resources depends on the specific resource and the nature of the resource affected. The time scale for sediment may range from five to 10 years for accumulation of sediment in road ditches to 100 years or more for accumulation of sediment in a reservoir. The time scale for a windstorm that reduces visibility on a highway is a single storm event with an infrequent return period of perhaps 10 years. The impact of fine sediment on stream ecology and dust in air causing respiratory problems is related both to single, infrequently recurring storms delivering a heavy concentration of fine particles and to multiple storms contributing to the long-term presence of fine particles in the water and air.

Whereas both erosion and its on-site impacts occur at the same location, the off-site damage caused by sediment may extend a great distance from the sediment source. Identifying the upstream sediment source for an off-site impact can be very difficult, although watershed hydrologic and water-quality models such as AGNPS and SWAT can be used to help link upstream sediment sources with downstream sediment yield (Arnold et al., 1998; Bosch et al., 1998). A watershed-management plan should focus on the source areas causing the greatest contribution to the sediment yield. A model can be used to examine alternative watershed-management plans to determine the plan that best meets the sediment-control objectives. A common regulatory approach, especially for small areas, is to require all land users in the sediment-producing areas to apply BMPs rather than to treat specific areas.

TECHNICAL TOOLS FOR CONSERVATION PLANNING

Conservation planning most often is performed at the local and site levels to select practices to control erosion and sediment delivery from specific sites. The technical tools for local conservation planning include erosion-prediction technology, soil-loss tolerance, land classification, land-use maps, soil survey data, topographic maps and digital elevation data, aerial photographs, and technical specifications for conservation practices. Erosion-prediction technology is discussed in Chapter 5, soil-loss tolerance was discussed earlier in this chapter, and erosion-control practices are discussed in Chapter 7. Several other tools for conservation planning are described briefly here.

Socioeconomic Information

Soil conservation practices must fit local socioeconomic conditions. For example, a conservation practice that requires mechanical power is not acceptable when only animal and human power is available. The practice must fit site conditions. For example, a conservation practice based on farm equipment designed for large fields is not appropriate for small, irregularly shaped fields in mountainous areas. A sediment control impoundment that takes considerable land out of production is inappropriate where land ownership exists in very small units. A conservation practice that requires a large capital investment, such as the purchase of a new machine, is not acceptable for subsistence farming. Religion is sometimes a consideration. For example, certain religions do not allow the use of tractors. Considerable management skills are required to apply and maintain some conservation-tillage systems. The local population may abandon conservation practices almost immediately after installation because the local farmers cannot or will not maintain them (Heusch, 1981; Shaxson, 1981).

Information as to land-unit boundaries, land use, and personal knowledge of local customs is essential. Modern remote sensing and geographical information systems (GISs) greatly facilitate data collection, assembly, processing, analysis, and presentation of information. For example, land-use information can be overlain with topographic and soil data to determine patterns of land use and the relations between land use and other land characteristics.

Land Capability Maps and Soil Survey Data

Land and soil vary in their capability to support various uses: A proposed conservation practice must fit the land and soil characteristics where it is applied. For example, steep lands with shallow, gravelly soils may not be suited for cultivation at all and should only be used for pasture or forest. Land capability classes have been developed so that land could be mapped according to land-capability class (USDA, NRCS, 2001d):

- *Class 1 soils* have slight limitations that restrict their use.
- *Class 2 soils* have moderate limitations that reduce the choice of plants or require moderate conservation practices.
- *Class 3 soils* have severe limitations that reduce the choice of plants or require special conservation practices, or both.
- *Class 4 soils* have very severe limitations that restrict the choice of plants or require very careful management, or both.
- *Class 5 soils* have little or no hazard of erosion but have other limitations, impractical to remove, that limit their use mainly to pasture, range, forestland, or wildlife food and cover.
- *Class 6 soils* have severe limitations that make them generally unsuited to cultivation and that limit their use mainly to pasture, range, forestland, or wildlife food and cover.
- *Class 7 soils* have very severe limitations that make them unsuited to cultivation and that restrict their use mainly to grazing, forestland, or wildlife.
- *Class 8 soils* and miscellaneous areas have limitations that preclude their use for commercial plant production and limit their use to recreation, wildlife, or water supply or for aesthetic purposes.

Soil and topographic characteristics mainly determine land-capability classes. Soils vary in their properties over the landscape, but soil-mapping units can be defined and the soil properties within each unit determined. Figure 8.2 and Table 8.1 illustrate a soil survey map and a partial description of soil properties for soil-mapping units represented on the map. The soil survey description gives the T value and other important data concerning soil properties related to erodibility, runoff, infiltration, plant growth, construction, and waste disposal that determine land suitability for various land uses.

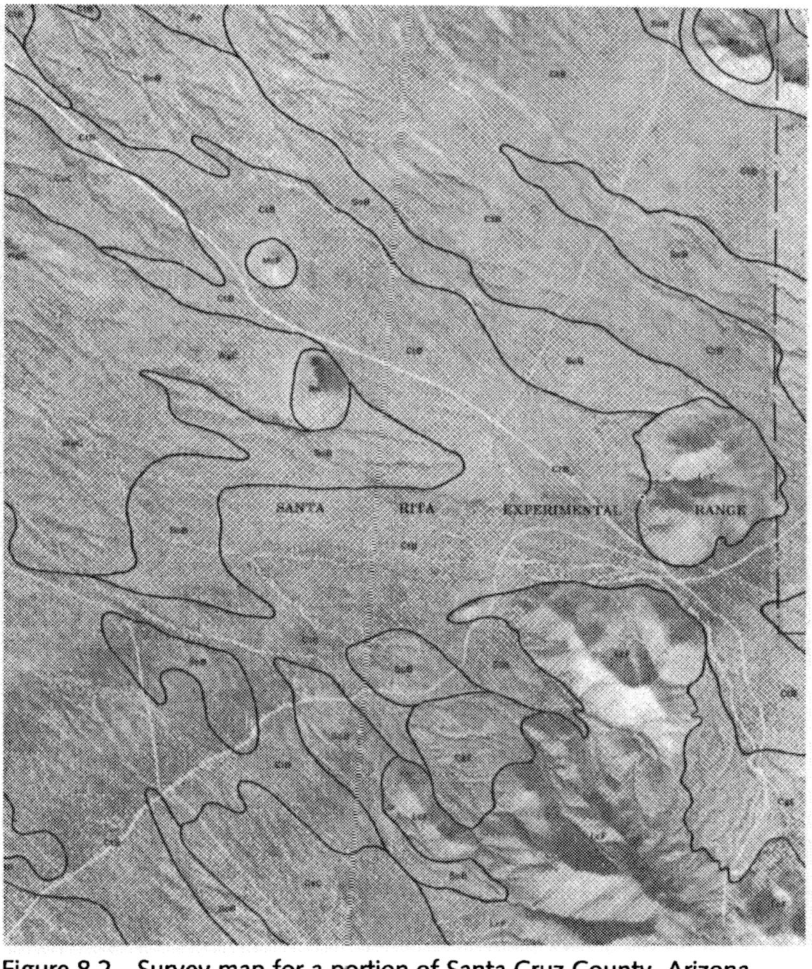

Figure 8.2 Survey map for a portion of Santa Cruz County, Arizona. (From USDA, SCS 1979.)

Table 8.1 Selected Data for Surface Layer on Soils in Santa Cruz County, Arizona

Map Symbol	*Soil Series*	*Slope Steepness (%)*	*USDA Texture*	*Runoff Potential*
CtB	Comoro	0–5	Gravelly sandy loam	Low to moderate
SoB	Sonoita	1–8	Gravelly sandy loam	Low to moderate
WhC	White House	1–15	Clay loam	Moderate to high
An	Anthony	1–3	Fine sandy loam	Low to moderate

Source: Data from USDA, SCS (1979).

Topographic Data

The site typically is visited to determine the current land use and topographic information needed to develop a conservation plan for the specific site. Land-use maps may not be current, and most of the readily available topographic maps are at too small a scale (1/24,000 or smaller) to accurately determine slope length, steepness, and shape values for erosion-prediction technologies. However, topographic and land-use maps, including those in digital form, are used in watershed and regional conservation planning. Watershed boundaries and sediment delivery paths are delineated on topographic maps and digital elevation data using with topographic analytical tools, such as TOPAZ. Hydrologic models, such as AGNPS and SWAT, to compute erosion and sediment movement through large watershed systems (Bingner et al., 1997).

Practice Specifications

The proper selection, installation, and maintenance of conservation practices require technical specifications that are frequently available from conservation agencies (USDA, NRCS, 2001b). Also, this information is frequently available from universities and other educational and technical training centers. Installing and maintaining practices according to the specifications helps to ensure that the practices will perform well. Table 8.2 illustrates USDA's specifications for contouring.

Table 8.2 Partial Specifications for Contour Farming

Contour Farming

Definition

Tillage, planting, and other farming operations performed on or near the contour of the field slope.

General criteria applicable to all purposes

Minimum Row Grade. Row grades for soils with slow to very slow infiltration rates or for crops sensitive to ponded water conditions for periods of less than 48 hours are designed with positive row drainage of not less than 0.2% on slopes where ponding is a concern.

Maximum row grade. The row grade shall be aligned as closely as possible to the contour to achieve the greatest erosion reduction. The maximum grade of rows must not exceed 2% or one-half of the up-and-down hill slope percent used for erosion prediction, whichever is less. Up to 3% row grade may be permitted within 150 ft (50 m) of the approach to a grassed waterway, field border, or other stable outlet.

Minimum ridge height. The ridge height shall be designed to reduce soil erosion compared to that of rows oriented up and down the slope. As a minimum, this practice shall be designed to achieve a ridge height of 0.5 to 2 in. (10 to 50 mm) during the rotation period that is most vulnerable to soil erosion.

Source: Adapted from USDA, NRCS (2001b).

LOCAL SOIL CONSERVATION PLANNING FOR ON-SITE EROSION AND SEDIMENT CONTROL

Owners and users of private lands routinely make land-use management decisions that affect soil erosion. The extent to which these decisions are consistent with conservation objectives depends largely on the decision maker's conservation ethic, which directly relates to his or her commitment to conservation, knowledge of soil-conservation practices, and willingness and ability to install and maintain conservation practices. Formal conservation planning is unlikely without participation in public programs, such as a cost-sharing incentive-based program or regulations requiring a permit to undertake land-disturbing activities. Although formal conservation planning may not occur, the decision makers often are influenced by their awareness for a need to care for the land (land stewardship), the importance of soil conservation, and sources of technical information. Public organizations, such as extension and conservation agencies, and trade publications are important information sources, and professional societies and private suppliers of erosion-control products hold conferences and workshops, and publish information concerning soil conservation practices. Public incentive and regulatory programs typically require development of a formal conservation plan before funding or a permit is released.

The major principles of conservation are (1) erosion and sediment are controlled to applicable standards; (2) the landowner's/user's preferred land use is accommodated if possible; (3) the recommended conservation practice is profitable, convenient, maintainable, and considers personal preferences; (4) local customs are respected; (5) resources to install and maintain the recommended practice are available; and (6) the resulting land-use system is sustainable over the long term.

The following examples illustrate how conservation planning occurs.

Example 1: Cost Sharing for Agricultural Land

A particular government program makes public money available to assist farmers with soil conservation. The program requires that a formal conservation plan be developed and that practices be installed and maintained according to specifications. Farmers become aware of the program from neighbors, field days, or the news media and arrange for a visit with the local office of the conservation agency responsible for the program.

A conservation planner typically visits the site and brings along a soil survey and a topographic map. The site is located on the soil survey map, as in Figure 8.2, and the soil-mapping units with erosion problems are identified in the field. The farmer describes the preferred land use, and a list of alternative conservation practices are developed by the farmer and the conservation planner. The objectives of the conservation plan are iden-

Figure 8.3 Sediment transport from field to drainage system. (Courtesy of USDA, NRCS.)

tified, which in this example are erosion control to avoid land degradation and sediment control from the fields to protect an adjacent stream channel (Figure 8.3).

The field where the conservation practice or practices are to be installed is inspected and the one-fourth to one-third most-erodible part of the field is identified, which serves as the basis for the conservation plan. Using the most-erodible part of the field in planning is conservative because erosion is reduced more than necessary for most of the field. An alternative planning criterion could be one where the plan is based on comparing average soil loss for the field to soil-loss tolerance. In a few situations, the most erodible part of the field can be managed differently from the remainder of the field, but in most cases, farmers prefer to apply the same practice to the entire field. *Prescription farming technology* is becoming available where farming operations can be varied as soil properties vary over the landscape, but this technology cannot accommodate the wide range of conditions necessary to control erosion equally throughout the entire field. The planning criterion usually depends on the requirements of the public program providing funds and the policy of the conservation agency administering the program.

The planner has information for the major factors that affect erosion,

Table 8.3 Alternative Practices for a Conservation Plan near Columbia, Missouri[a]

Practice	Soil Loss	
	tons / acre	*metric tons / ha*
Clean till, continuous corn, up and downhill	15	33
Mulch till, continuous corn, contoured	5	11
No-till, continuous, corn, contoured	0.3	0.8
Rotation of clean till corn, corn, hay, hay, hay, contoured	4	10

[a] Field with a hillslope length of 150 ft (46 m) and 6% steepness and a soil loss tolerance of 5 tons/acre (11 metric tons/ha) per year.

including climate for the location, soil erodibility as described in the soil survey data, and the topographic information collected in the field itself. The fourth major factor that determines erosion rates is land use, so the purpose of this conservation plan is to choose a land-use and management system that will control interrill and rill (or wind) erosion to a rate less than the T value. An erosion-prediction technology is used to estimate soil loss for several proposed land-use alternatives, and the ones that give an estimated soil loss less than the T value are considered acceptable. A typical example of conservation-practice alternatives is shown in Table 8.3. The current practice of clean-tilled continuous corn is too erodible at 15 tons/acre (33 metric tons/ha) per year. Therefore, a different cropping-management system is needed to reduce soil loss to a T value of 5 tons/acre (11 metric tons/ha) per year for this particular soil. The farmer and planner together develop a list of alternative conservation practices (Chapter 7) and use the erosion-prediction technology (Chapter 5) to estimate soil loss for each practice. The practices that give estimated soil losses that are less than soil-loss tolerance are considered acceptable. The farmer chooses a practice from among this group that best fits his or her interests, including profitability, convenience for the farming operations, and personal preference. For example, the no-till practice is superior for controlling erosion, but the farmer is concerned about weed control and chooses the mulch–till system because weed control might be less difficult. The corn–corn–hay–hay–hay rotation also provides acceptable erosion control, but the farmer can make more profit with continuous corn and thus does not choose this rotation.

The second part of the conservation plan is control of the off-site sediment delivery. Previous regional water-quality planning on the watershed scale has established criteria for sediment delivery from farm fields to streams in the watershed. Based on the erosion prediction results, the planner determines that while the selected erosion-control practices meets the soil-loss tolerance criteria, it does not meet the criteria for sediment delivery into the stream adjacent to this field. Thus, the farmer must apply a sediment-control practice in addition to the erosion-control practice. Two major possibilities are considered: (1) a system of small impoundments near the outlet of the concentrated-flow channels that drain the field, and (2) a filter strip system of dense grass along the stream channel. Application of erosion-prediction technology shows that either system will provide the required sediment control. Both the filter strip and the impoundments are estimated to reduce sediment yield to 1.0 ton/acre (2.2 metric tons/ha) per year. The farmer chooses the filter-strip system because of reduced costs, although maintenance, such as mowing and periodic reseeding, will be required. The farmer likes the additional wildlife habitat provided by the filter strip, and public funds are available for installing the filter strip but currently are not available for installing impoundments.

The planner assists the farmer in installing the conservation system selected and then returns to the farm periodically to ensure that the practices are maintained according to specifications. Percent ground cover is a major variable in determining the effectiveness of conservation-tillage systems, such as the mulch–till system selected by the farmer. This conservation plan requires maintaining an average annual 35% residue cover after planting, which will be assessed by the planner as a part of the compliance process.

Example 2: Construction Permit Required for Construction Site

In this situation, an area of about 200 acres (80 ha) will be extensively disturbed during development. A local ordinance, based on federal law, requires an erosion-control plan before a permit is issued allowing land disturbance to begin. The local ordinance is most concerned about sediment leaving the construction site and increased runoff rates after development, that will cause downstream channel erosion. The developer hires a consulting engineer to develop the erosion-control plan. Some form of sediment control is required because erosion control by itself will not control sediment delivery from the site, especially for those periods when construction leaves the land exposed and vulnerable to erosion. Further, the development dramatically increases runoff from the site and will cause downstream gully and channel erosion that continues after development. Small retention basins that contain storm runoff are used to reduce peak runoff rates so that downstream channel erosion will not occur. These basins serve as sediment-control basins during construction, but the sediment trapped in the basins during construction is removed before the basins are placed in service as runoff retention basins after development is complete. Therefore, a part of the erosion-control plan is spreading and stabilizing the sediment removed from the basin after construction is complete.

The erosion-control plan requires that construction occur in phases so that the land exposed to erosion at any one time is minimized. Also, land disturbance is timed to avoid the most erosive periods of the year to the extent possible. The BMPs of temporary mulch and vegetative cover minimizes soil-exposure time, and permanent vegetation is established as soon as possible following completion of the project (Chapter 7). Strips of vegetation are left as long as possible along concentrated-flow areas to trap sediment. Another component of the erosion-control plan is surface runoff management. Once concentrated-flow areas are disturbed, they are stabilized promptly using manufactured "rolled" erosion-control blankets. A fabric filter fence is installed around the perimeter of the site to prevent sediment transported by overland flow from leaving the site.

A critical part of the plan is to ensure that all practices are properly installed because poor installation defeats the best of practices. Other critical parts of the plan are frequent inspection and prompt repair. This carefully developed and executed erosion- and sediment-control plan allows construction activities to proceed without delay and does not require an extended effort to fix on- and off-site erosion problems during the years after development is complete and before all affected areas were finally stabilized.

Example 3: State Regulations Require an Erosion- and Sediment-Control Plan for a Landfill

Modern landfills involve large, intensively disturbed areas with almost 100% soil exposure during the filling phase. Therefore, the erosion-control plan for this landfill requires that the landfill be opened and filled in phases to minimize the area that is disturbed and exposed to erosion at any one time. The plan also requires wind-erosion control to prevent blowing dust and control of sediment delivery by runoff during filling. As each phase is completed, the plan requires that the area be quickly graded, a soil cap placed on the surface, and vegetation established. Vegetation establishment is difficult on this landfill, as it is on many landfills, because the soil quality of the cap is poor. The revegetation plan requires selection of vegetation species that are suitable for the adverse soil properties, local climate, and a low level of maintenance before the vegetation becomes well established. A quick-growing cover crop is seeded along with the permanent vegetation to control erosion quickly and facilitate establishment of the permanent vegetation. Fertilizers, lime, and micronutrients are applied at seeding. The design criterion is that the permanent vegetation is to control the interrill and rill erosion rate to less than the T value used for a comparable agricultural soil.

The erosion-control plan includes a surface-drainage plan to transport runoff from the site without causing excessive rill or gully erosion. The landfill has a large upper flat surface that produces large amounts of runoff. A diversion is placed around the outer edge of this area to collect this runoff and convey it to a protected channel that transports the runoff down the sideslope of the landfill in a downdrain (Figure 7.4). The possibility of placing diversions and terraces on the sideslopes was considered but rejected because the diversions could "overtop," causing gully erosion that would expose the underlying waste materials. The drainage plan takes into account differential settlement caused by the decomposition of organic material buried in the landfill. Extra freeboard (depth) is added to the drainage channels. The erosion- and water-control plan requires that the erosion- and sediment-control practices be inspected periodically and repaired promptly if failure occurs.

Example 4: State Regulations Require an Erosion- and Sediment-Control Plan as Part of the Reclamation Plan for a Surface-Mined Area

Surface mining drastically disturbs vegetation, soil, and topography. In the early days of surface mining, before the 1940s, the land frequently was left in the topographic configuration created by dumping spoil material, which is waste rock and other material overlying the resource being mined, without shaping the spoil material into any particular topographic shape. The spoil material often did not support vegetation, eroded rapidly with downstream transport of sediment and pollutants that degraded water quality, and was usually considered an eyesore. Through time, legislation and regulations evolved that require reclamation of surface-mined lands to the approximate original contour prior to mining and a land-use appropriate for the region. To ensure adequate reclamation, including erosion control, of lands surface-mined for coal, regulatory agencies hold a financial bond for five years in humid climate regions, where vegetation is readily established, and for 10 years in drier regions, where vegetation is more difficult to establish. The land reclamation process is a form of land conservation.

The first step in reclamation is removal and stockpiling of topsoil before mining begins. After mining, the spoil is graded to the approximate original contours (AOC) of the land prior to mining. The stockpiled topsoil is placed over the graded spoil material and vegetation is established (Chapter 7). Erosion is controlled using temporary mulch and vegetation until the permanent vegetation is sufficiently established to control erosion. The mulch usually is anchored to the soil by crimping. Fertilizers and other soil amendments are applied to facilitate seed germination and plant growth. Supporting practices, including ripping and contouring, are used to increase infiltration that provides the soil moisture needed for plant growth. The permanent vegetation on many reclaimed sites consists of native species, especially in rangeland regions, and hay and pasture grasses in areas with sufficient rainfall. Where the land before mining was prime cropland, regulations require that the land be returned to cropland after reclamation and that the production level be comparable to that before mining.

The topography is the foundation for reclamation and erosion control (Toy and Black, 2000). Hillslopes on this site are shaped in accordance with geomorphic principles, and a water-conveyance system is designed to accommodate the runoff and sediment loads immediately after reclamation. The topography is designed realizing that the landscape will evolve through time. Complex hillslope profiles, with an upper part that is convex and a lower part that is concave and where the upper convex portion accounts for about one-third and the lower concave portion accounts for about two-thirds of the hillslope length, are chosen. The concave portion of these hillslopes will induce deposition, reduce sediment delivery, and reduce the likelihood of gullies beginning at an overfall into a channel and migrating into the hillslope sides. The three-dimensional configuration of the landscape, including the drainage network, is designed to control the location of the head of concentration-flow channels so that these channels will not erode too far into the hollows. Shaping the landscape is a compromise because the cost of moving spoil and soil to achieve a particular topographic shape is a major cost in reclamation.

Preferred topographic shapes are not always the least expensive to construct.

Diversions are placed on the long hillslopes to shorten slope length to prevent excessive rill erosion that could lead to gully erosion. These diversion channels are carefully designed and installed so as not to erode or to cause deposition. Extra capacity is added to prevent overtopping in case deposition occurs. Establishing sustainable vegetation at the available management level is the single most important factor in controlling long-term erosion on reclaimed land. Most reclaimed sites in the semiarid and arid rangeland regions are revegetated with native species and require minimal long-term maintenance. Through time, a plant community develops that is a function of local climate, soil properties, and grazing pressure from cattle and wildlife. In regions where rainfall is plentiful and the reclaimed land is used as pastures and to produce hay, maintenance can be much more intense, with routine and cost-effective applications of fertilizers and other soil amendments.

CONSERVATION PLANNING BY GOVERNMENTAL UNITS

The major objectives of conservation planning by governmental units are to develop and implement public policy to minimize excessive erosion. Public policy affects both public and private lands. Public lands come under the jurisdiction of an oversight governmental agency and must be managed according to the specific intent and requirements of legislation and regulation. From time to time, new legislation is passed that requires changes in agency management of public lands. Although the discussion below focuses on public policy for private lands, much of the discussion also applies to conservation planning on public lands.

The spatial scale considered in conservation planning and the development of public policy is determined by the level and jurisdiction of the governmental unit developing the policy. For example, the spatial scale of conservation planning by a city is limited to the region over which the city has jurisdiction. Conservation planning and development of public policy occurs at national, regional, state, county, city, and local government levels. Each government level has certain powers to raise funds by taxation and to pass, implement, and enforce legislation related to conservation of the land under its jurisdiction. Often, programs operate as a partnership between national, state, and local government levels.

Conservation analyses are also carried out on the global scale by both private and public organizations. The Food and Agriculture Organization (FAO) of the United Nations is one public organization involved in global conservation. The FAO has conducted erosion inventories in various countries throughout the world and has published technical documents concerning erosion control for specific regions (Dudal, 1981). Private organizations such as the World Bank and the World Watch Institute have analyzed erosion worldwide and reported on the severity and seriousness of erosion problems (Brown and Kane, 1994). The World Bank, USAID (U.S. Agency for International Development), numerous developed countries, and nonprofit foundations provided both technical and economic aid to developing countries to facilitate soil conservation efforts (USAID, 2001; World Bank, 2001).

Several global-scale assessments of soil degradation by erosion and other processes have been made (Hurni, 2000; Pimental et al., 1987; 1995; Zachar, 1982). These assessments typically are based on very limited data, many assumptions, and much data extrapolation. Not surprisingly, these assessments are intensely debated (Boardman, 1998; Crosson, 1995; Trimble and Crosson, 2000). The purpose in this section is not to engage in the debate but to describe erosion assessment and the development and implementation of a public soil conservation policy.

Erosion Inventory

The first step is to inventory erosion and estimate its impact. The first decision that must be made is how to inventory erosion, whether erosion is measured directly in the field, measured indirectly through sediment-yield measurements, or estimated using erosion-prediction technology. In Chapter 6 we discussed various techniques for measuring erosion; in Chapter 4, requirements for using sediment delivery measurements for estimating soil loss; and in Chapter 5, erosion-prediction technology. Although no single method is perfect for conducting all erosion inventories, use of erosion-prediction technology at many sample points is preferred overall for developing public policy involving large, diverse geographic regions (USDA, NRCS, 1997b).

One of the major requirements of an erosion inventory is that erosion and its impact be estimated at many sample points rather than assuming average conditions for a large area and then making the estimates. The best approach for an erosion inventory of interrill, rill, and wind erosion is to use an erosion-prediction model to estimate erosion for a set of sample points statistically selected to provide the desired accuracy and to allow data to be aggregated in the various forms needed to develop public policy and serve other purposes. The erosion rate is estimated at each sample point using climate, soil, topographic, and land-use data collected at the sample point.

Information on the spatial variability of upland erosion is needed to identify local areas with severe erosion problems so that these areas can be treated with erosion-control practices. Also, erosion rates are used in crop-yield models to estimate deterioration of crop yields because of erosion. The mathematical relationship between crop yield and erosion and the spatial variability of both crop yield and erosion are nonlinear. Therefore, an average erosion rate for a large area is a poor indicator of the impact of erosion on crop yield. The proper approach is to determine the impact of erosion on small areas and weight the effect on each area by the fraction of the total area that each subarea represents to determine the overall impact of erosion (Arndt et al., 2001; Griffin et al., 1988; Perrens et al., 1985).

The USDA has used this approach since the 1970s to conduct periodic inventories of interrill, rill, and wind erosion on private lands across the United States (USDA, NRCS, 1997b). In cooperation with Iowa State University, the USDA identified about 800,000 sample points on U.S. nonfederal land, which provided accuracy of soil-erosion estimates to the county level, an area of about 30 miles by 30 miles (50 km by 50 km). The estimate of average annual erosion rate at each sample point were aggregated in a variety of ways, including Figure 1.5, which shows the areas in the United States where erosion is considered to be excessive. Also, an estimate of the erosion impact on crop yields at each sample point was made based on the soil conditions at each point, and those impacts also were aggregated in different ways, such as the effect of erosion on crop yield by region. The data from an erosion inventory can be analyzed to determine how erosion varies by region, land use, crop, major soil type, farming method, and other major variables. By repeating the inventory every five years, the USDA can assess changes in erosion patterns through time and has determined, for example, that erosion has decreased by about one-third from 1982 to 1997. The periodic USDA erosion inventories provided the essential data needed to develop and pass the 1985 Food Security Act, which was landmark soil conservation legislation. The inventories identified the locations and extent of highly erodible land and cropland that was eroding at an excessive rate. The inventory data were used to estimate the program costs of alternative proposals, which was information needed by the legislators. Subsequent inventories provided the data to evaluate the progress and the success of the legislation.

Developing and Implementing Public Soil Conservation Policy

Successful public soil conservation policies fit the local cultural, social, economic, and political systems (Hudson, 1981b; Shaxson, 1981). Although the United States has been an international leader in developing public soil conservation policy, U.S. policy cannot be taken and applied directly in other countries. The public soil conservation policy of any country must fit the unique circumstances of that country. However, a portion of the U.S. experience is presented as a case study.

In the United States, much of the land, especially cropland, where erosion rates are high, is in private ownership. The government cannot simply order improved soil conservation on private lands because of private ownership rights. Because of these private property rights and traditions, U.S. public soil conservation policy for private lands since the 1930s has been based on voluntary, cost-sharing, and incentive-based programs (Napier and Napier, 2000). By the early 1980s, policymakers had begun seeking ways to increase the effectiveness of public soil conservation programs (Esseks et al., 2000; Weber and Margheim, 2000). With previous programs, allocated funds were not always spent where soil conservation was needed most. The far-reaching 1985 Food Security Act was passed based on support by a coalition of traditional soil conservationists, environmentalists, and farm groups (Napier, 1990). Each group supported the legislation because their special interests were met. To soil conservationists, the legislation would result in the application of soil-conservation practices to far more land than had occurred with any previous legislation. To environmentalists, the legislation would reduce non-point-source pollution from agricultural lands, which is a major source of non-point-source pollution. To farm groups, the legislation would provide financial aid to farmers at a time when the farm economy was very depressed. The legislation was designed to: (1) remove the most erodible land from crop production, and (2) require conservation practices on the land remaining in production where erosion was excessive (Napier, 1990). It also directed (targeted) public funds to the lands with the most severe erosion problems.

The government "rented" specific parcels of land identified as highly erodible, which removed these lands from cropping and kept them under protective vegetative cover for the 10-year period of the agreement (Napier, 1990). In exchange, the landowner and user received a cash payment from the government for these lands, known as *conservation reserve program* (CRP) *lands*. Lands were designated as highly erodible based on estimated erosion rates under nonconservation farming practices compared to the soil-loss tolerance value for the soil of the land (Napier and Napier, 2000). This ratio of the erosion rate to the soil-loss tolerance rate is an erosion index that takes into account the climatic erosivity of the location, soil, topography, and the extent to which erosion could be reduced without special assistance. That is, if soil conservation practices could be applied to the land without excessive costs, the land would not be considered as highly erodible. Farm groups liked the CRP program because it removed land from production, which raised farm income by providing a direct cash payment and by increasing agricultural commodity prices.

The other component of the program was to reduce excessive erosion on lands that remained in production. Although the program remained voluntary, like previous programs, this legislation significantly increased the incentive to participate in soil conservation. An increased incentive was needed to greatly expand the area treated with conservation practices. Farmers were required to apply conservation practices to the land as a condition for participation in other government programs that provided price supports for agricultural commodities. Given the poor agricultural economy in the late 1980s, income from the commodity price-support programs was almost a necessity. This incentive was so strong that some farmers felt that the public soil conservation program for cropland had shifted from a voluntary to a mandatory and quasiregulatory program. The program also seemed regulatory because a formal conservation plan was required. USDA personnel inspected fields to determine farmers who could be penalized "if found out of compliance."

This landmark public program greatly accelerated the development and adoption of conservation tillage, a new erosion-control technology, by the private sector (Chapter 7). The benefits of conservation tillage extend beyond soil conservation by increasing the profitability of farming at a time of economic difficulty for farmers. The best incentive for soil conservation is a conservation technology that has obvious and immediate benefits, especially profitability. The short-term requirements of annual farming expenses, loan payments, and simply remaining in business often override the long-term benefits of soil conservation. The benefits of conservation tillage caused this erosion-control practice to be used long after this particular government program ended in 1995, which results in a permanent decrease in soil erosion. Subsequent programs have been put in place that continue the directions of the successful 1985 Food Security Act.

LESSONS FROM THE U.S. CONSERVATION MOVEMENT

The overall state of soil and land in the United States today can be compared to that of the early twentieth century. During the past 70 years, the public investment in soil conservation has been estimated to be $100 billion. Television documentary films show scenes of wind erosion so severe during the Dust Bowl days of the 1930s that houses were abandoned because they filled with windblown sediment (Figure 8.4). A mass migration of the rural population from Oklahoma to California occurred during this period, described in the novel *The Grapes of Wrath,* partly because of wind erosion and a depressed national economy (Steinbeck, 1939). No such mass migration has occurred since then, and houses filled with windblown sediment are unthinkable today. Pictures that showed severely

Figure 8.4 Wind erosion of cropland in Kansas, 1935. (Courtesy of USDA, NRCS.)

eroded hillslopes were common in the literature before the 1930s showing severely eroded hillslopes (Figure 8.1), but few such hillslopes are evident today. Without question, U.S. farmland is in far better condition today than a century ago, despite assessments that consider a major portion of U.S. cropland to be severely degraded (Hurni, 2000; Pimental et al., 1995). Certainly, U.S. farmland is not of the same quality as before it was brought into crop production beginning in the late eighteenth century and which expanded through the early twentieth century. Farmland is in much better condition today that it was a half-century ago based on the authors' personal experience. Great progress has indeed been made in U.S. soil conservation.

Lands mined by surface methods and reclaimed since 1977 are in far better condition than the lands surface-mined before the 1940s, when spoil often was left where it was dumped. Water quality in streams draining these areas is greatly improved. Fabric fences that trap and retain sediment on construction sites are common today. Before the 1970s, the runoff very likely would have transported and deposited that sediment in streets and other off-site locations, causing problems. Forestry practices on private lands have changed from the intensive, clear-cut practices that were common a half-century ago that allowed excessive erosion. Logging-road design, construction, maintenance, and reclamation after logging have improved to reduce substantially the sediment delivered to forest-land streams in many cases. Improved land conservation is the result of many factors contributing in combination. In agriculture, yields for some crops have increased dramatically because of advanced agricultural science and technology. The increase in yield alone reduced soil erosion by about 60% on midwestern U.S. fields devoted to continuous row crop production. The same technologies greatly reduced the likelihood of total crop failure, so land rarely is left completely bare and exposed to erosive forces. A major reason for the disastrous wind erosion during the Dust Bowl was a severe drought that caused crop failures and left insufficient vegetation cover to protect the land. Greatly increased crop yields have allowed highly erodible land to be retired from crop production and placed in permanent cover. Economics, rather than soil conservation policy, brought about this change, because farming poor-quality, highly erodible lands is not competitive with farming high-quality, erosion-resistant lands.

Another key factor was that off-farm employment was available, so a livelihood could be earned from sources other than subsistence farming of poor-quality land. Steep hillslopes that were farmed in the early twentieth century were converted to pastures and forests several decades ago. Much of the southeastern United States that was in highly erodible cotton production in the early twentieth century is now forested, resulting in almost no erosion. Cropping systems have changed dramatically in U.S. agriculture, where large areas are now in continuous row cropping rather than crop rotations that included hay and pasture that would have been common a half-century ago. The shift away from these crop rotations tended to increase erosion, but modern conservation-tillage technology has reduced erosion below the rates of the 1950s while increasing profitability and competitiveness in national and international commodities markets. In the 1950s, a plowed field would have been almost bare of crop residue from last year's crop. Today, a residue cover of 20% or more is common, and many fields are "no-till" farmed with residue covers of 50% or more.

Public policy has definitely played a major role in land conservation on all U.S. lands. The extensive erosion-research programs that began in the 1930's would not have occurred without public funding. Erosion and land degradation was a crisis situation in the United States by the 1920s and something had to be done. The creation and activities of the Natural Resources Conservation Service and the Cooperative Extension Service in nearly every county across the United States has provided soil-conservation education and technical assistance to nearly every corner of the country. The landmark legislation of 1985 that resulted in significantly decreased erosion was the product of innovative legislation and public policy. An exciting consequence was the development of new conservation technology, primarily by the private sector, that will continue to be used because of its inherent benefits.

The increase in soil conservation on all U.S. lands, including mining and construction lands, parallels increased soil conservation on agricultural lands. The NRCS has been a major contributor in providing technology and technical assistance for the conservation of these lands. Another major factor has been major regulatory legislation from the local to federal level that has required erosion control, sediment control, and reclamation of disturbed lands. Federal agencies, such as the U.S. Department of Agriculture, Forest Service; U.S. Department of Interior, Bureau of Land Management and National Park Service; and partner state agencies have played a major role in the management of public lands in the United States. The U.S. Department of Interior, Office of Surface Mining, Reclamation and Enforcement, and partner state agencies have played a major role in land reclamation following surface mining. The U.S. Environment Protection Agency, along with comparable state agencies, has been the lead agency for protecting off-site water and air quality. The U.S. Department of Interior, Geological Survey, has provided scientific support to these federal and state agencies as well as to the private sector (e.g., Shown et al., 1981; Staubitz and Sobashinski, 1983). These agencies and the policies that they implemented occurred because land conservation has the political support of the people.

The debates over preservation of natural resources versus utilization will continue appropriately. Although the effectiveness, value, burden, and cost of public programs will continue to be debated, the fundamental perspective of the U.S. public is one of environmental stewardship and a willingness to commit public funds to environmental protection, including land preservation and soil conservation on land used for a variety of purposes where land-disturbing activities must occur.

SUGGESTED READINGS

Agassi, M. (ed.). 1996. *Soil Erosion, Conservation, and Rehabilitation*. Marcel Dekker, New York.

El Swaify, S. A. and D. S. Yakowitz (eds.). 1998. *Multiple Objective Decision Making for Land, Water, and Environmental Management*. Davis Publishers, New York.

Kebede, T., and H. Hurni (eds.). 1992. *Soil Conservation for Survival*. Soil and Water Conservation Society, Ankeny, IA.

Kral, D. M. (ed.). 1982. *Determinants of Soil Loss Tolerance*. ASA Special Publication 45. American Society of Agronomy, Madison, WI.

Lal, R., and F. J. Pierce. 1991. *Soil Management for Sustainability*. Soil and Water Conservation Society, Ankeny, IA.

Larson, W. E., G. R. Foster, R. R. Allmaras, and C. M. Smith (eds.). 1990. *Proc. Soil Erosion and Productivity Workshop*. University of Minnesota, St. Paul, MN.

Napier, T. L., S. M. Napier, and J. Turdan (eds). 2000. *Soil and Water Conservation Policies and Programs: Sucesses and Failures*. Soil and Water Conservation Society, Anlceny, IA.

Schmidt, B. L., R. R. Allmaras, J. V. Mannering, and R. I. Papendick (eds.). 1982. *Determinants of Soil Loss Tolerance*. American Society of Agronomy, Madison, WI.

9

Perspectives and the Future

Soil erosion continually shapes and reshapes the land. Geologic erosion rates are a natural part of landscape development. The land and its soil must be used by the human population for the production of food, fiber, shelter, and fuel. The activities associated with the production of these commodities disturb the land and accelerate soil erosion rates. Although land use and accelerated erosion seem necessary and inevitable, our challenge is to prevent land degradation through the application of conservation and reclamation practices. There is always a competition between land use and preservation.

The public and scientific interest in soil erosion seems to be cyclic, waxing and waning through time. Although concern for the environment has grown dramatically in recent years at global, national, and local scales, protecting the land from excessive soil erosion does not enjoy the celebrity and political stature afforded other environmental concerns. The present interest in soil erosion seems to focus on the off-site protection of air and water quality rather than protection of the land itself. Controlling sediment delivery by controlling erosion, however, protects both on- and off-site resources. In some cases, sediment delivery can be controlled while accelerated soil erosion still degrades the land.

The importance of soil conservation for present and future generations cannot be overemphasized. In simple terms, the long-term existence of humanity depends on controlling water and wind erosion. Soil conservation, indeed enriching the soil resource, is an ongoing task, not a "fix and forget" endeavor. This job falls to those directly connected to the land, the millions of farmers, conservationists, reclamationists, government agency personnel, and research scientists around the world who work with the land and soil day by day.

This chapter concludes our soil erosion textbook with a summary of the essential lessons to be taken from the preceding chapters and a discussion of the factors that are likely to affect soil erosion and conservation in the near future. Based on our life experiences in working on and with the land and after our review of material to prepare the book, we offer our personal perspective on soil erosion in this chapter.

ESSENTIAL LESSONS

Reflecting back, you realize that each chapter contains a large amount of information. Erosion is a complex topic. The essential lessons in each chapter are summarized below. These are the concepts and principles that we believe you should always remember. We review these essential lessons for emphasis.

- *Chapter 1: Introduction.* Erosion is a major, perhaps *the* major, environmental problem worldwide. Soil can be destroyed much faster than it is formed.
- *Chapter 2: Primary Factors Influencing Soil Erosion.* Several interrelated factors control erosion processes. These factor vary temporally and spatially.
- *Chapter 3: Types of Erosion.* Water erosion is examined in a spatial context: interrill, rill, concentrated flow area, gully, and channel. Wind erosion is examined based on transportation mode: creep, saltation, and suspension.
- *Chapter 4: Erosion Processes.* Sediment load is controlled by either the amount of sediment produced by detachment or by the transport capacity of the water or air flow. Deposition occurs where sediment load exceeds transport capacity.
- *Chapter 5: Erosion-Prediction Technology.* Models are used to estimate erosion rates for various purposes. The purpose, data requirements, and available resources determine the model choice.
- *Chapter 6: Erosion Measurement.* Erosion-measurement tech-

(continues)

(*continued*)

niques vary in accuracy and cost. The measurement purpose, requirements, and resources determine the technique choice.

- *Chapter 7: Erosion and Sediment Control.* Erosion-control practices reduce detachment or transportation capacity, or both. Practices exist to control erosion in nearly all situations. Proper installation and maintenance of practices are necessary.
- *Chapter 8: Land Conservation.* Conservation involves both the public and private sectors. The public sector supports conservation, while the private sector implements conservation practices.
- *Chapter 9: Perspectives and the Future.* Soil erosion is an environmental problem that must be solved for future generations. Soil degradation may increase or decrease in the future, depending on the interplay of several factors.

When viewed from temporal and spatial perspectives, and from physical, economic, and social perspectives, soil erosion is arguably *the* major environmental problem worldwide. The fact is that soil erosion accelerated by human activities can destroy soil much faster than it can form, rapidly depleting the soil resource. The cost of erosion and sedimentation is measured in billions of dollars annually.

Climate, soil, topography, and land use, especially vegetative cover and soil management, determine the character of erosion processes. These factors influence water and wind erosion processes in different ways. The factors, and their constituent subfactors, interact among themselves. The factors vary temporally and spatially; hence, erosion varies temporally and spatially as well.

Erosion and sedimentation includes a group of processes that detach, entrain, transport, and deposit soil and other earth materials. A practical approach to erosion research and control examines water erosion in a spatial context, including interrill, rill, concentrated flow, ephemeral gully, permanent gully, and stream-channel erosion. Wind erosion is examined on the basis of the sediment-transport modes of creep, saltation, and suspension.

The amount of sediment produced by water or wind erosion is controlled by detachment processes (detachment limited) or by the transport capacity of the water or air flow (transport limited). The total erosivity of water or air flow is divided between sediment detachment and transport. Deposition occurs when the sediment load exceeds the transport capacity.

Erosion-prediction technologies (models) are developed for various purposes. These models may be simple or complex, requiring few or many data inputs. The model use commonly determines the model choice within data-input requirements and available-resource constraints.

Several techniques are available for measuring erosion that vary in accuracy, equipment, and personnel requirements, and hence operating costs. As accuracy increases, costs usually increase. The measurement purpose commonly determines the technique choice within data requirement and resource constraints. The most accurate and expensive measurement technique is not always necessary to serve the measurement purpose.

Erosion-control practices must reduce particle detachment or reduce the water- or air-flow transport capacity, or both. Great progress has been made in developing and refining erosion-control practices. The technology exists to control erosion on nearly all disturbed sites, ranging from simple mulches and revegetation to manufactured products with design features for specific applications. Proper installation and maintenance are assumed for all erosion-control practices.

Conservation usually involves a partnership between the public and private sectors. The public sector supports conservation through government agency activities and financial incentive programs. The private sector implements and maintains the conservation practices. Conservation planning tools, maps, models, and practice specifications are available through public agencies.

Several factors suggest that soil conservation will improve in the future. Several other factors suggest that soil degradation will continue and may even worsen, at least in some parts of the world. Excessive soil erosion is an environmental problem that must be solved.

FUTURE FOR SOIL CONSERVATION

Our hope and expectation is that soil conservation will become a high priority worldwide in the years to come. At the same time, we are apprehensive that soil erosion could increase in the future, especially in areas of the world where the land use is intensive and conservation resources are limited.

Hope for Soil Conservation

The renewed conservation ethic, conservation-technology development, expanding technology-transfer conduits, and an effective public-policy approach suggest that soil conservation will improve and spread during the coming years in those countries with adequate economic resources and without overwhelming production pressures on the land. Whether by genetic coding or experience through the ages, the human species generally seems to possess an attachment to the land manifested in a sense of stewardship, even for those who do not have a direct connection to the land. There are structural evidences of soil conservation from many civilizations, as discussed in Chapter 1. By 1769, George Washington was experimenting with conservation farming practices at Mount Vernon. The Virginia school of experimental agriculturalists, including Thomas Jefferson, devoted considerable attention to the soil-erosion problem. Patrick Henry, soon after the American Revolution, is said to have declared that "since the achievement of independence, he is the greatest patriot who stops the most gullies" (Bennett and Lowdermilk, 1938). Conservationists are among good company.

Presently, the conservation movement is growing as a part of the overall interest in the environment. Many international organizations, national governments, and nongovernment organizations (NGOs) embrace environmental protection of air, water, and land. The national commitment to conservation is expressed in laws, regulations, and support programs. A common philosophy is that if the land cannot be conserved or reclaimed satisfactorily, it should not be disturbed. Although much of the current interest in soil erosion focuses on the off-site impacts of sediment, this philosophy and commitment is conducive to development and spread of soil conservation throughout the world.

Those using the land are subject to public pressures from many directions to minimize on- and off-site damages and to conserve or reclaim disturbed lands. Large corporations place greater value on environmental protection than ever before, perhaps because of this public pressure and the importance of positive public relations. Corporations see environmental protection as a social responsibility, or at least a part of doing business. Land stewardship is good business practice. A society expresses its environmental expectations through public policy with economic incentives and disincentives.

Decades of scientific research throughout the world provide the foundation for conservation strategies. Great strides have been made in our understanding of erosion processes. This knowledge is incorporated into the erosion-prediction technologies discussed in Chapter 5. These models are used to prepare efficient conservation plans by comparing the effectiveness of various erosion-control practices.

A wide variety of erosion-control practices are available to the conservationist and reclamationist for cropland, rangeland, mine site, and construction lands. New erosion-control practices, adaptations, and new equipment become available each year. Some are the products of agricultural and erosion-control industry research. Some result from design modifications following field applications. Sometimes, practices result from government incentive programs. The private sector develops new products and erosion-control techniques to address the requirements of legislation. The best public policies for soil conservation result in the development and implementation of effective practices that endure after the policy and its programs expire. A worldwide research infrastructure exists to address new erosion problems and adapt erosion-control practices to local conditions.

New agricultural science and technology that increase productivity can be considered conservation practices in that they allow the demands of the marketplace to be satisfied using fewer acres (hectares) of land. Today, each cropland acre (hectare) in the United States produces nearly three times as much as produced on the same acre (hectare) in 1935 (USDA), NRCS, 1997a). Highly erosive lands can be retired from production and placed in complete grass cover to minimize erosion rates, as with the Conservation Reserve Program (CRP). Removing highly erosive lands from production and increasing productivity from the lands least susceptible to erosion is a very effective conservation strategy. To remove land from production, however, displaces people and alternatives for earning a living must be available. Even for those that remain on the land, soil conservation requires resources, at least in the short term. A vibrant economy facilitates soil conservation.

Although agricultural science has its risks and certainly should be pursued very carefully, we must continually increase land productivity while maintaining long-term sustainability. Mechanization has not been the evil as it is sometimes portrayed. A very strong case can be made that mechanization actually reduced soil erosion because erodible lands were taken from production that could not be farmed with these methods.

There is ample evidence that a partnership between the public and private sectors can succeed in reducing erosion rates. Since 1982, cropland erosion decreased from 3.1 billion tons to 1.9 billion tons per year (2.8 to 1.7 billion metric tons) per year (USDA, NRCS, 1997b, revised 2000). Figure 9.1 shows the decline in erosion rates during the 15 year period. Today, it is rare to see the deeply rilled, gullied, and windblown agricultural lands of the 1930s. Reclamation is required by law for mine lands and erosion and sediment control are required by law for construction lands.

Clearly, there has been major progress in the continuing fight to control erosion on disturbed lands. Regardless of the work yet to be done, the United States is not experiencing the crisis that it was in the 1930s with the severe erosion problems and degraded lands of that time. We should be proud of our progress, but not complacent. Rains will continue to fall, winds will continue to blow, and lands will continue to be disturbed. Climate change could exacerbate soil-erosion problems during the next century, but our forecast is that science, technology, and industry again will provide solutions to these problems.

In the early days of surface mining, topsoil often was not salvaged for subsequent use in land reclamation. Today, topsoil salvage is required.

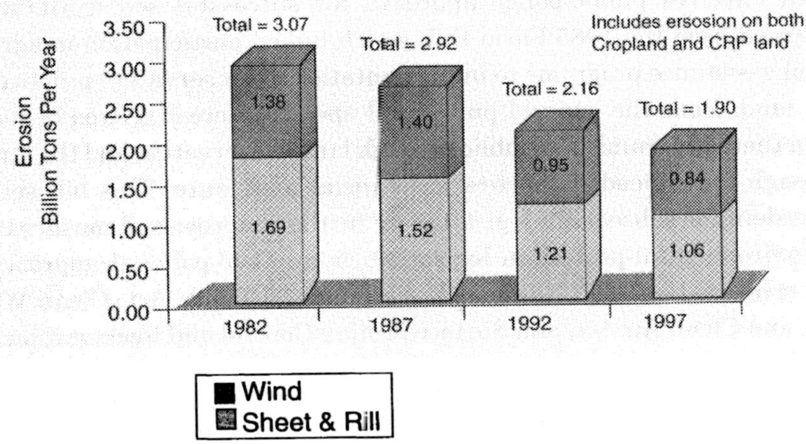

Figure 9.1 Decreasing erosion rates for cropland and CRP land in the United States, 1982–1997. (Courtesy of USDA, NRCS.)

During the past 20 years, techniques were developed to construct plant-growth media (mine soil, soil substitute) from available geologic materials, mulches and compost, and biosolids (sewage sludge). Although not a replacement for natural topsoils, these materials permit revegetation, accelerate soil development, and initiate nutrient recycling (Haering et al., 2000).

Technology transfer is the means by which conservation technologies become available to the users. Government agencies and professional associations publish manuals describing numerous erosion-control practices and their specifications, as noted in Chapters 6 and 7. For agricultural lands, government agencies at the federal, state, and local levels assist farmers in preparing conservation plans tailored to their lands. For mining and construction lands, meetings, workshops, and trade shows are held annually by professional organizations where new and improved erosion-control practices are presented and discussed. Scientific journals present the results of experiments conducted to develop and test new erosion-control practices. Trade magazines describe field experiences with various erosion-control practices.

Electronic technology-transfer methods are emerging rapidly. Research reports, bibliographies, erosion-prediction technologies, manuals and guides, meeting announcements, and chat lines are available in offices or homes at any hour of the day. Appendix C provides a partial list of erosion-related Web sites from which others can be discovered.

An effective public-policy approach for successful soil conservation emerged from the 1985 Farm Bill, which linked participation in agricultural assistance programs to implementation of conservation practices on the land. Both the general public and special-interest groups benefited from these programs, the public through land conservation and the farmer through much-needed short-term financial assistance. This bill set the precedent for subsequent legislation to further improve soil conservation.

Environmental-protection legislation is another political approach to soil conservation. The National Environmental Policy Act, Clean Water Act, and Clean Air Act, and Surface Mining Control and Reclamation Act, are among the statutes that directly or indirectly seek to control erosion and sedimentation. Prior to these laws, sediment and other pollutants from disturbed lands commonly degraded air and water quality. Under present law, such environmental impacts are illegal and can result in fines amounting to thousands of dollars per day.

Apprehensions Regarding Soil Conservation

Despite our faith that science and technology will find solutions to future soil erosion problems, there remain apprehensions. Population growth, poor economic conditions in parts of the world, the capacity for land disturbance, and ineffective political approaches suggest that soil erosion could increase in the future. The projected world population now exceeds 6 billion people and is expected to exceed 9 billion by 2050 according to the U.S. Bureau of the Census [www.census.gov/ (2001)], (Figure 9.2). The world is adding residents at the rate of 146 per minute.

Population growth is the primary threat to environmental quality. Each person requires food, fiber, shelter, minerals, and fuel and can be expected to do whatever is necessary to meet those requirements. Survival is instinctive. In addition, most people expect increases in their standards of living during their lifetime. As population increases, more resources are needed to meet basic demands and provide a few amenities. The extractive industries that produce these resources always cause some measure of land degradation, including accelerated erosion, unless effective conservation or reclamation practices are implemented.

A closer look at population growth reveals a very grim prospect for the soil. Ninety-nine percent of the global natural increase in population (difference between births and deaths) now occurs in the "less developed" countries of Africa, Asia, and Latin America (U.S. Bureau of the Census,

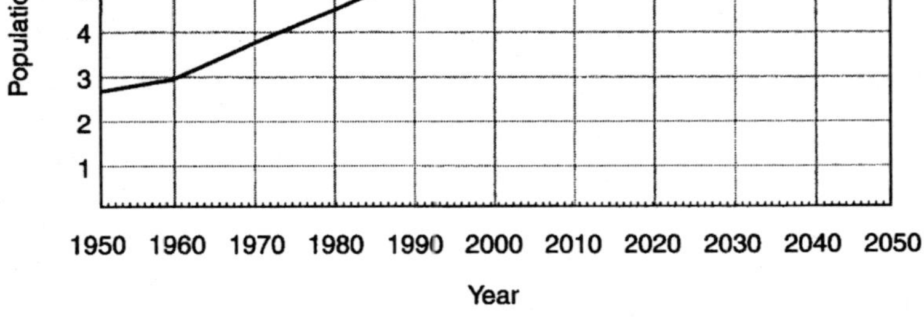

Figure 9.2 World population growth, 1950–2050. (From U.S. Bureau of the Census, 2001.)

[www.census.gov/2001.] The period of development in any country is likely to be a period of exploitation (Bennett and Lowdermilk, 1938). These countries usually do not have the technology and economic capability to conserve the land.

Poor economic conditions effectively can prohibit conservation and erosion control. Not everyone has the opportunity to be a steward to the land, as discussed in Chapter 8. Survival, literally and economically, often requires short-term decisions. It is not unusual for the adoption of a soil conservation system to entail some reduction in immediate cash return (Bennett and Lowdermilk, 1938). Frequently, an agricultural practice, such as conservation tillage, proves to increase profits while conserving the soil. These practices can be the foundation of practical soil conservation if adaptable to conditions in less developed countries.

Sometimes, land is just not suitable for a particular use, such as agriculture or grazing. Profits are low in most years, losses occur in some years, erosion rates are high, and conservation is not affordable. Abandoning the land is possible only if there are jobs elsewhere. Conservation requires discretionary income, at the personal and national levels. A true challenge for the future is to find ways to meet the basic food requirements of the growing population or to find ways to redistribute the available food supply.

While technological advancements contribute to the hope for improved soil conservation in the future, other technological advancements will contribute to land degradation. For example, heavy equipment used in the field, forest, mine, and at the construction site, becomes larger, sturdier, and faster every year. Processing equipment improves the profitability of resource exploitation. Transportation, seismic, and global-position equipment facilitates resource exploration. With restrictive environmental legislation in "more developed" countries, mining, logging, and other extractive industries are moving to the less developed countries, those countries that desperately need the economic benefits provided by these industries and therefore may be somewhat more tolerant of land degradation and soil erosion.

In the past, ineffective public-policy approaches achieved only limited success in controlling erosion. One of the common failings of public policy is that soil conservation does not continue after the policy and programs end. Various erosion-control practices can be researched and promoted, but until practices are discovered that have inherent benefits, especially profitability, the fight will continue in searching for ways to implement conservation on the land. Volunteer conservation programs tended to produce more planning than practice implementation and maintenance. Threats of fines and other penalties ring hollow when inspection and enforcement agencies are underfunded and understaffed. Significant progress has been made in the last 30 years (Figure 9.1), but we are still a long way from the goal of complete erosion control.

International agreements, national, state, and local legislation and regulation, and administrative systems, of themselves, do not solve erosion problems. The money has to be spent on the ground. Even in the United States, with the productivity of 1 of every 3 cropland acres (hectares) threatened by soil erosion and the productivity of 1 of every 5 rangeland acres (hectares) threatened by soil erosion, the public commitment for conservation on private land is well below the 1937 level, with less than one-half of the commitment made 60 years ago (USDA, NRCS, 1997a).

In the 1930s and 1940s, soil erosion was considered to be a national crisis in the United States. The scientists addressing water and wind erosion problems searched for understanding of erosion processes, but even more important was finding practical methods of erosion control. Substantial time was spent in the field observing erosion processes and their effects. There was an urgency to their mission. The knowledge that these scientists acquired with the available scientific technology remains impressive (e.g., Bennett and Lowdermilk, 1938). Today, the times have changed; soil erosion is no longer perceived to be a crisis, at least in the United States. The research emphases seem to be on erosion theory, processes, and modeling. Although great progress has been made in these aspects of erosion research, we are concerned that firsthand field experience has been neglected.

Despite the advances in our knowledge of erosion theory and processes over the years, erosion remains essentially an empirical science. High-quality databases are therefore absolutely essential for erosion studies. The data collection programs initiated during the 1930s to 1970s essentially ended in the 1970s as coordinated efforts directed toward erosion problems on agricultural lands. A modern database is needed, comparable to those used to develop the erosion-prediction technology in the 1950s but addressing contemporary crop varieties, agricultural-production methods, and conservation practices. Similarly, scientific research programs are needed to assess the effectiveness of manufactured erosion-control products.

There also seems to be a lack of decisive information concerning the damage caused by soil erosion. The available information ranges from broad generalities, based on numerous assumptions and data extrapolations, to localized case studies. There is not a good way that enjoys widespread acceptance to place a value on the land and then to determine the extent to which erosion degrades this value. Consequently, it is possible for some people to conclude that soil erosion is a major problem, whereas others conclude that it is no problem at all. The soil-loss tolerance concept is widely used as a criterion in conservation planning, but the scientific basis for particular values is very weak. It is a useful concept, but one that should be better defined.

These are the factors that suggest improvements in soil conservation or increases in soil erosion during the years to come. The actual direction depends on the interplay among these factors in any area. Although it is dangerous, perhaps foolish, to speculate about the future, we have mustered the courage to offer our perspectives. On a global scale, the severity of soil erosion probably will increase because of population pressures on the land in the less developed countries, poor economic conditions, technology transfer difficulties, and the hesitancy of the more developed countries to provide adequate soil-conservation assistance. The less developed countries will continue to have difficulty weighing short-term necessities against long-term benefits. For the more developed countries, priorities appear to be the real obstacle to soil conservation. People in less developed countries are likely to experience high and possibly increasing erosion rates on their land because they cannot afford erosion-control practices. People in more developed countries should experience lower and possibly decreasing soil-erosion rates on their land because the techniques and incentives are in place for effective erosion control.

CONCLUSIONS

Accelerated soil erosion is a problem that must be solved to ensure the welfare of future generations. The scientific foundation, evaluation techniques, and erosion-control practices exist to reduce erosion rates substantially in nearly all situations. The motivation and opportunity to implement soil conservation programs frequently are the pieces needed to complete the puzzle.

Although erosion and erosion-control research during the past 70 years provides a solid understanding of erosion processes and a variety of erosion-control practices, much more remains to be done. Changing technologies present new challenges and offer new possibilities. The erosion-prediction technologies will expand to include additional erosion-control practices, especially those used on mined and construction lands. In the near future, erosion-prediction technologies will be coupled with geographic information systems with the accuracy necessary for effective conservation planning and erosion control. Erosion-measurement techniques will continue to incorporate highly accurate electronic measurement, data-logging, and display techniques. Existing erosion-control practices will be made more economical and adapted to the environmental conditions of the less developed countries. Hopefully, the more developed countries that export the technologies for land disturbance also will export the technologies for soil conservation.

Now, can you imagine yourself as a part of the solution to the soil-erosion problem? There is a world that needs trained soil-erosion scientists and conservationists. There are possibilities in the public and private sectors for those who understand soil erosion processes, prediction, measurement, and control.

Appendix A: Soils

Soil properties affect soil erosion and, in turn, soil erosion affects soil properties. The purpose of this appendix is to provide basic information concerning soil properties and their relationship to erosion processes. Soil is the upper layer for much of the Earth's surface. Natural soils are the product of physical weathering, chemical weathering, and biological processes operating on geologic materials through time. The weathering processes alter geologic material to produce the mineral components of soil. The life cycles of plants and animals and excretory processes of animals add organic matter to the soil. Various organisms live within and interact with the mixture of mineral and organic components.

Human activities influence both soil formation and soil erosion. Land uses that change the shape, hydrology, and vegetation cover of the land also change the processes of soil formation. Soils can be "manufactured" for use as plant-growth media on severely disturbed lands, such as mining and construction sites, by combining available soil, rock fragments, organic matter (compost, manure, sewage sludge), and other amendments. Land uses also change erosion rates and alter soil properties. For example, removal of vegetation cover accelerates wind erosion rates, and wind erosion selectively removes fine particles from the soil surface, causing the proportion of sand-sized particles to increase through time. Soil properties, as modified by human activities, influence future land-use opportunities.

SOIL PROPERTIES

The response of soils to erosive forces is determined by the interaction of physical and chemical properties. Basic soil properties that affect erosion processes include soil texture, structure, chemistry, and organic-matter content. These properties largely determine soil structure, aggregation, permeability, profile development, and soil erodibility. Soil properties,

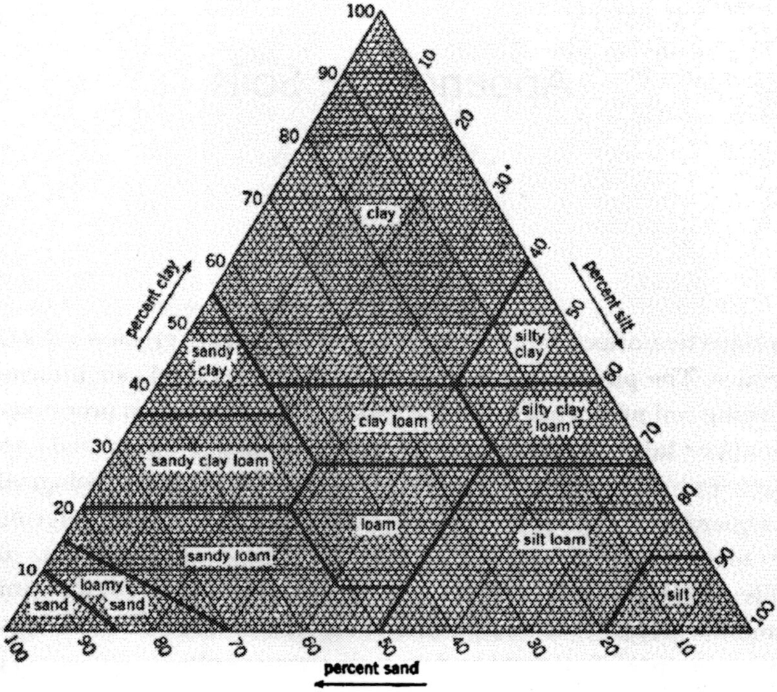

Figure A.1 Percentages of sand, silt, and clay in the basic soil textural classes. (Courtesy of USDA, NRCS.)

erosion processes, and depositional environments determine sediment characteristics.

Soil Texture

Soil texture refers to the distribution of primary particle sizes making up the soil. Soil texture is perhaps the single most important soil property because texture influences soil functions in many ways. Particle-size groups, or *separates*, are classified on the basis of particle diameter as sand [coarse to fine, 2.0 to 0.05 mm], silt [0.05 to 0.002 mm], or clay [0.002 mm], according to the U.S. Department of Agriculture system. There are other classification systems that vary slightly in particle-size ranges. The percentages of the separates can be plotted on the soil texture triangle (Figure A.1) to determine the appropriate textural descriptor for a soil. For example, a soil consisting of 33% each of sand, silt, and clay is referred to as a *clay loam*. Soil texture influences the ability of a soil to absorb and hold water for plant use. Clays commonly restrict the entry of water into the soil. Sandy soils have little surface area on which water molecules can be stored. Hence, medium-textured soils, such as silt-loams, provide the best moisture conditions for plant growth. Particles larger than (2.0 mm) are classified as gravels, and those greater than 5.0 mm usually are considered rock fragments for soil-loss estimation.

Organic Matter

In addition to mineral particles, soils contain varying amounts of organic matter. Plants and animals contribute organic matter to the soil while decomposition processes within the soil alter the physical and chemical characteristics of organic matter. The amount of organic matter in the soil is determined by (1) climate conditions, (2) nature of the vegetation, (3) topography, (4) soil texture, and (5) drainage conditions (Foth and Turk, 1972). Organic matter is a very important and highly reactive soil constituent that recycles plant nutrients, provides surfaces for water storage, and binds mineral soil particles into larger aggregates that resist wind and water erosion.

Soil Structure

Soil structure refers to the arrangement of soil particles and aggregates. Structural types include *granular, platy, blocky,* and *massive soils*. Structure influences the movement of air and water through the soil. For example, granular soils allow much more water to enter the soil than do massive soils. For a given rainfall event, a surface composed of granular soils produces much less runoff than a surface composed of massive soils. As water or wind moves across the soil surface, the attractive forces among soil particles that produce the platy, blocky, and massive structures must be overcome by the shearing forces of water or wind in order to detach particles from the larger mass and initiate the erosion process.

Soil Aggregation

Soil particles commonly are clustered into a single mass, or *aggregate*, by the attractive forces among individual particles. As the percentage of clay, organic matter, iron, and aluminum oxides in the soil increase, the proportion of the soil in aggregates and the size of the aggregates increase. Soil aggregation leaves pore spaces between the aggregates through which air and water move, increases infiltration, reduces crusting, and reduces the susceptibility of the soil to wind and water erosion.

Soil Permeability

Soil permeability refers to the ease with which air and water can move into and through the soil. Soil texture, structure, and aggregation determine soil permeability. The process of water moving into the soil is known as *infiltration*, so as soil permeability increases, the infiltration rate increases. As infiltration increases, runoff production and erosion decrease (Appendix B).

Soil Chemistry

Soils also possess chemical properties that greatly affect erosion processes and plant growth, including soil-solution reaction (pH), electrical conductivity, sodicity, salinity, clay mineralogy, and the presence of iron and aluminum oxides. Soil-solution reaction refers to the amount of free hydrogen ions in the soil, ranging on the pH scale from below 7 for acidic soils, to above 7 for alkaline soils. Electrical conductivity (EC) measures the concentration of ionized constituents in the soil solution. Conductivity greater than 4 mmhos/cm indicates a saline soil. Exchangeable sodium percentage (ESP) measures the sodicity of the soil solution. An ESP value of 10 or more indicates a sodic soil. High or low pH, high electrical conductivity, high salinity, or high sodicity may prevent plant growth, leaving the soil surface exposed to erosive forces. In addition, sodic soils tend to disperse when wetted, resulting in poor structural stability and greater vulnerability to erosive forces. Iron and aluminum oxides are binding agents, prevalent in subsoils, that increase the soil's ability to resist erosive forces.

Clay Mineralogy

Clay mineralogy is the product of chemical weathering processes operating on geologic materials under particular environmental conditions through time. There are two major groups of clay minerals, the silicate clays and the oxide clays. The silicate clays are found throughout the world, whereas the oxide clays are found in humid tropical regions. The basic building blocks of the silicate clays are sheets of silicon–oxygen tetrahedrons and sheets of aluminum octahedrons. Combinations of these structures and their specific chemical composition produce various types of clay minerals with various properties (Birkeland, 1999) that affect erosion processes directly and indirectly.

Clay minerals differ in the amount of expansion and contraction during wetting and drying. Water between layers of clay minerals and between clay particles cause the shrink–swell, plastic, and cohesive characteristics of clay-rich soils. A study by Young and Mutchler (1977) showed that as the percentage of montmorillonite clay in the soil increased, the soil erodibility decreased.

Clay minerals possess very high surface area per unit mass. For example, a fine sand may have a surface area of 100 cm^2/g, while a clay may have a surface area of 100,000 cm^2/g. Clays have the ability to store large quantities of water adsorbed to these surfaces to support plant growth. Pollutants, however, also may adsorb to these clay surfaces and move downstream with clay sediments. The amount of surface area varies by clay type.

Soil Profiles

Soil-forming processes usually create distinct soil layers, called *horizons*. The *A-horizon* is the surface layer of soil, sometimes referred to as *topsoil*, where most of the plant roots, organic matter, and biological activity occurs in the soil. Erosion processes strip away the fertile A-horizon, reducing soil depth and the water-storage capacity of the soil. Soluble and very fine particles are translocated from the A-horizon to the underlying *B-horizon* by water percolating through the soil.

Colloidal clay, fine-size organic matter, and oxides of iron and aluminum accumulate in the B-horizon, sometimes referred to as the *subsoil*. The B-horizon soils usually, but not always, are less fertile than the A-horizon soils. This is one reason why removal of the A-horizon by erosion results in decreasing agricultural productivity without soil management, such as applications of fertilizers and organic matter.

Tillage mixes soil horizons and incorporates organic matter through the soil to the tillage depth. This changes the physical and chemical properties of the soil and, as a result, the response of the soil to erosive forces. Increasing organic matter in the soil increases the proportion of the soil in aggregates and the size of the aggregates.

Soil Erodibility

Soil erodibility refers to the inherent susceptibility of soil particles or aggregates to erosive forces. When rainfall energy, slope length and steepness, vegetation cover, and other factors are held constant, the erosion rates for some soils are higher than the erosion rates for other soils. Differences in soil erodibility result from differences in physical and chemical properties through complex interactions. There have been many efforts to measure soil erodibility (Bryan, 1977). One common method estimates a soil's resistance to water erosion on the basis of particle-size distribution, organic-matter content, soil structure, and soil permeability. Soil strength is a useful measure of a soil's resistance to wind erosion.

Temporal and Spatial Variability

Soil properties vary with the annual changes in weather conditions that bring wetting and drying, and freezing and thawing, of soils. Organic matter is added to the soil and decomposed during the year. Global climate changes may alter soil properties over the long term. Similarly, soil properties vary spatially, along hillslope profiles, and across the three-dimensional landscape. These changes in soil properties along a hillslope are recognized in soil classification as a soil *catena* or *toposequence*. These differences in soil properties result in differences in the hydrologic response to rainfall events and to the erosive forces of water and wind.

SEDIMENT PROPERTIES

Sediment is the product of erosion, consisting of primary soil particles and aggregates detached from the soil surface by raindrop impact, concentrated water flow, and wind. Sediment particles at the point of detachment tend to be larger than the primary particles composing the soil because of aggregation. Sediments composed mostly of clay-size material are referred to as *fine sediment* and those composed mostly of sand are referred to as *coarse sediments*. The sediment-size distribution produced by water erosion often is bimodal, with concentrations at diameters of about (30 and 200 μm) (Foster et al., 1985a). Sediment density, expressed as specific gravity, ranges from about 1.4 to 1.8 for aggregates to about 2.65 for primary particles.

Once in motion, sediment particles are carried along with the water and wind flow until the velocity and turbulence of the flow decrease and the particles respond to the force of gravity in the process of deposition. The fall velocity depends on the size and density of particles and the viscosity of the transporting flow, with the largest and densest particles falling fastest. Differences in fall velocities for sediment particles result in the stratification of sediment deposits. When the velocity and turbulence of the flow decrease, the largest particles are deposited first while the smaller particles are transported farther. For example, sediments deposited in water at the base of a hillslope are coarsest near the base of the hillslope and become progressively finer with distance from the base. Similarly, sediment deposited at a windbreak are coarsest near the windbreak and become progressively finer with distance from the windbreak.

SOURCES OF INFORMATION

The U.S. Department of Agriculture, Natural Resources Conservation Service (NRCS), has surveyed, mapped, and analyzed most soils throughout the United States. Additional information is available through the NRCS Web site (*www.nrcs.usda.gov*). Maps, profile descriptions, and a variety of tabulated data are contained in published soil surveys available through local NRCS offices. Soil conservationists, project planners, and reclamation personnel should obtain a copy of the survey for the area of interest.

SUGGESTED READINGS

Birkeland, P. W. 1999. *Soils and Geomorphology*. 3rd ed. Oxford University Press, New York.

Bryan, R. B. 1977. Assessment of soil erodibility: new approaches and directions. In: T. J. Toy (ed.), *Erosion: Research Techniques, Erodibility, and Sediment Delivery*, pp. 57–72. Geo Books, imprint of Geo Abstracts, Norwich, Norfolk, England.

Foster, G. R., R. A. Young, and W. H. Neibling. 1985. Sediment composition for nonpoint source pollution analyses. *Trans. ASAE* 28(1): 133–139, 146.

Foth, H. D., and L. M. Turk. 1972. *Fundamentals of Soil Science*. Wiley, New York.

U.S. Department of Agriculture, Soil Conservation Service, 1983. *National Soils Handbook*, no 430. U.S. Government Printing Office, Washington. D.C.

Young, R. A., and C. K. Mutchler. 1977. Erodibility of some Minnesota soils. *J. Soil Water Conserv.* 32:180–182.

Appendix B: Hydrology

Hydrology is the science of water. The purpose of this appendix is to provide basic information concerning hydrology and the relationships among hydrologic and erosion processes. The hydrologic cycle is the central concept in the study of water and refers to the continuous movement of water from the atmosphere to the Earth's surface, across the surface, and back into the atmosphere. Water moves by various hydrologic processes and may be stored temporarily. The major hydrologic processes involved in soil erosion by water are precipitation and surface runoff; the most important storage location is within the soil as soil moisture. Other important hydrologic processes are infiltration, percolation, evaporation, and transpiration. Other important storage locations are vegetation surfaces and depressions in soil surfaces. Soil moisture also reduces wind erosion by binding together soil particles.

PRECIPITATION PROCESS

Precipitation is the transfer of water from the atmosphere to the Earth's surface in the form of rainfall, snow, hail, or sleet. Rainfall amount and intensity drive soil erosion by water. Rainfall and temperature determine the decomposition rates of plant residue and litter on the soil surface and organic matter in the soil. For example, decomposition is much higher in the southeastern United States than in the northern Great Plains because of high rainfall and high temperatures. Plant residue and litter reduce the erosive forces of both rainfall and wind, while organic matter in the soil resists water and wind erosion by reducing soil erodibility, increasing soil aggregation, increasing soil moisture, and facilitating plant growth.

Rainfall varies temporally and spatially (Figures 2.1 and 2.3). Hence, erosion rates also vary temporally and spatially. It is important to consider these variabilities in scheduling soil-disturbing activities such as

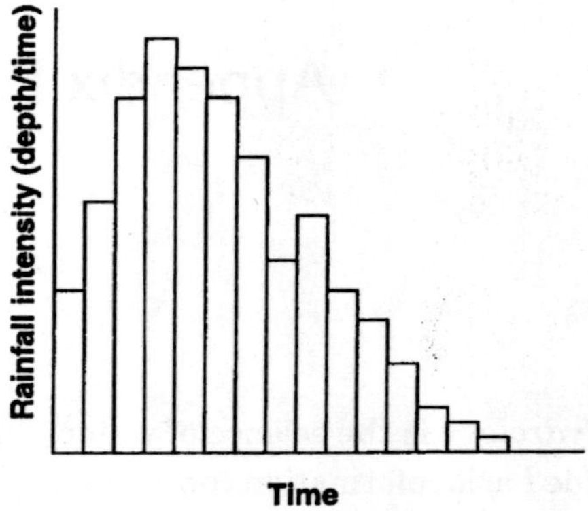

Figure B.1 Hyetograph showing the distribution of rainfall through time.

tillage and land-grading operations. Rainfall occurs as discrete events, and the distribution of rain through time is depicted in a hyetograph (Figure B.1). The two most important characteristics of a rainstorm, in relation to soil erosion, are the volume, measured as volume per unit area or depth (in. or mm), and the peak intensity, measured as volume per unit area per unit time (in./hr or mm/h).

High-magnitude rainstorms can cause very high erosion rates, and therefore such storms are important considerations in soil-erosion control. As the rainfall magnitude (including both amount and intensity) increases, the rainstorm frequency decreases. The frequency of a rainfall or runoff event can be expressed by a *recurrence interval* or *return period*. For example, the *10-year rainfall event* often is used in erosion- or sediment-control design and indicates that (1) an event of this magnitude has an occurrence probability of 0.10 every year, (2) in any year, there is a 10% chance that an event of this magnitude will be equaled or exceeded, and (3) *on average*, a rainstorm of this magnitude is expected once every 10 years (Dunne and Leopold, 1978). Thus, a 10-year rainstorm is smaller in magnitude than a 25-year rainstorm but occurs more frequently.

WATER STORAGE

The rainfall arriving at the Earth's surface may be stored temporarily on vegetation surfaces, in surface depressions, and in the soil. Storages can be described by the equation

$$I - O = \frac{\Delta S}{\Delta t} \tag{B.1}$$

where I in the inflow rate of water into a storage location, O the outflow rate of water from storage, and $\Delta S/\Delta t$ the rate of water storage change for that location. When the inflow rate is greater than the outflow rate ($I > O$), water accumulates in storage ($\Delta S/\Delta t > 0$). When the inflow rate is less than the outflow rate, the water in storage decreases through time.

Most storage locations function like a leaky bucket being filled from a garden hose. When the holes are above the bottom of the bucket, all the water that initially flows into the bucket is retained. However, as the level of the water reaches the height of the first holes, water begins to flow out of the bucket, but the bucket continues to fill as long as the inflow rate from the hose is greater than the outflow rate from the holes. When the bucket becomes completely full and water flows over the edge of the bucket, the outflow rate becomes equal to the inflow rate and storage in the bucket is completely filled. The holes in the bucket can be so large and the water outflow so great that the water level never reaches the top of the bucket. The weight of water above the holes (hydraulic head) creates the pressure at the holes. As the height of water above the hole increases, the pressure and flow velocity through the holes increase.

When the water inflow from the hose stops, water continues to drain through the holes in the bucket and the water stored in the bucket begins to decrease. The outflow rate decreases as the weight of water above the holes decreases. Some water remains in storage below the holes in the bucket.

Interception by Vegetation

Rainfall may be intercepted by vegetation cover before reaching the soil surface. Very early in a rainfall event, all of the intercepted rainfall is retained on plant leaves and stems, as when all the water is retained in the bucket, as described above. When the rainfall duration and intensity are low, almost all the rainfall can be retained on the vegetation and little reaches the soil surface. However, as rainfall continues or the intensity increases, the water-storage capacity of vegetation surfaces is filled, as when the bucket is full, and any additional water drips from the vegetation or flows along plant surfaces to the soil. The total amount of water that reaches the soil surface is the total amount of rainfall minus the amount retained on the vegetation. The amount of water intercepted and stored on vegetation is a function of the type, density, and growth stage of the vegetation.

Depression Storage

Rainfall received directly at the surface and falling or flowing from vegetation may be stored in small depressions created by (1) tillage and earth-moving machinery, (2) animal and human traffic; (3) plant stems, roots, and litter; and (4) deposition of wind-transported sediment. Generally, the soil that forms these depressions is permeable, so the depressions are leaky, as with the bucket described above. Water does not accumulate in the depressions until the filling rate exceeds the leakage rate of water flowing into the soil. Thus, rainfall intensity must exceed the leakage rate before the depressions begin to fill. The flow from depression storage as leakage begins as soon as water enters the depression, but most flow from depression storage occurs when the depression storage is filled. The weight of water in the depression creates a hydraulic head above the soil that drives water into the soil, while water in excess of the depression's storage capacity spills from the depression until the outflow rate equals the inflow rate, as with the bucket.

The volume of depression storage is a function of the number and size of the depressions (surface roughness). Surface roughness largely is a function of soil properties and the mechanisms that disturb the soil. Roughness decreases through time as raindrop impacts erode the ridges around the perimeter of the depression and deposit the sediment within the depression. When the soil surface is very rough, runoff can be greatly reduced or eliminated by depression storage.

INFILTRATION PROCESS

Infiltration is the movement of surface water into the soil, expressed in units of volume per unit area per unit of time or depth per unit of time. The infiltration rate is limited by the rate at which water is available to the soil or by the maximum rate at which water can enter into the soil as determined by soil properties. The maximum infiltration rate is known as the *infiltration capacity* of the soil. When rainfall occurs at a uniform rate, the infiltration capacity decreases (Figure B.2). Rainfall intensity, however, varies through time during almost all storms, so the actual infiltration rates in the field increase and decrease (Figure B.2). The shape of the infiltration curve is a function of the temporal variability of rainfall intensity and the temporal and spatial variability of soil properties that affect the infiltration process.

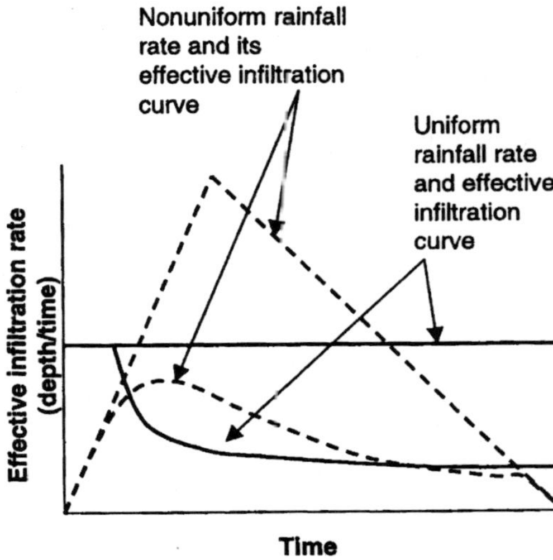

Figure B.2 Curves showing the change in infiltration through time for uniform and varying rainfall. (Flanagan et al., 1988.)

The infiltration rate is a function of the size and connections among soil pore spaces (soil permeability) as well as the extent to which the pores already are filled with water (antecedent soil moisture). Until the pores are filled, flow in the soil is referred to as *unsaturated flow*, but when the pores become filled, the flow is referred to as *saturated flow*. The infiltration rate reaches a minimum when the soil becomes saturated and this rate of water movement is referred to as the *saturated hydraulic conductivity* of the soil. After a rainstorm, gravity causes soil water to drain from the pore spaces. The soil water drains to a moisture content known as *field capacity*. Thereafter, evaporation and water uptake by plants further reduce soil moisture until the soil-moisture content reaches the *wilting point*, a condition where water is so tightly attached to soil particles that plants cannot remove any more water from the soil. The soil-moisture content between field capacity and wilting point is known as *available soil water*, available for plant use (Foth and Turk, 1972).

Most soils are layered with distinct horizons (Appendix A) and the horizon with the slowest hydraulic conductivity can control the infiltration rate at the soil surface. Also, raindrops impacting the soil surface can cause a thin [about 0.04 to 0.12 in.] compacted layer to develop on the soil surface. This dense layer, known as a *soil seal* when wet and a *soil crust* when dry, greatly restricts infiltration and controls the infiltration rate at the surface. Thus, the rate and amount of infiltration for a storm event depends on the soil moisture at the beginning of a storm (antecedent soil moisture) and the presence of a soil seal, in addition to the physical and chemical properties of the soil. A seal can greatly increase water erosion by reducing infiltration and increasing runoff, while a crust can greatly reduce wind erosion by increasing surface resistance.

RUNOFF PROCESS

When the rainfall intensity exceeds the infiltration capacity of the soil and the interception and depression storages are filled, the excess water begins to flow downslope. The reduction in runoff is due *largely* to infiltration but is *partly due* to water storage on the soil surface. Runoff during a storm event varies through time because of the variation in rainfall intensity and infiltration. The peak runoff rate is reduced and delayed in relation to the peak rainfall intensity. The rate of water storage as a flow is a function of (1) overland-flow slope length, (2) slope and channel steepness, (3) hydraulic roughness on the hillslopes and in the channels, (4) the size of the watershed, and (5) the excess rainfall rate (rainfall intensity minus infiltration rate). Hydraulic roughness is the *drag* that the surface exerts on runoff. Hydraulic roughness also increases as the roughness of the soil surface and plant stems, residue, and litter increase to slow the runoff. On very short (3 ft, 1 m) hillslopes, such as those of interrill areas, the water-storage rate is negligible, so that the runoff rate equals rainfall intensity minus infiltration rate. However, on longer (150 ft, 50 m) overland-flow hillslopes, the water-storage rate is significant and must be considered in infiltration and runoff computations (Skaggs et al., 1969).

The peak runoff rate also is reduced and lags in time behind the peak rainfall intensity as watershed size increases because of water storage and the travel time of water moving down the hillslope and through the stream channel network. Hydrographs show the variation in runoff through time (Figure B.3). Both runoff amount and peak runoff rate are used as measures of runoff erosivity for estimating water erosion.

EVAPORATION AND TRANSPIRATION PROCESSES

Evaporation and transpiration are processes of water transfer from the soil and plants into the atmosphere. Evaporation of water from the soil, vegetation surfaces, or depressions occurs when water molecules acquire

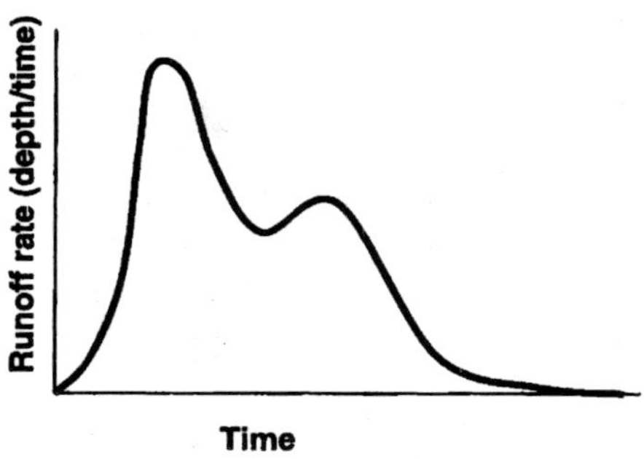

Figure B.3 Hydrograph showing the distribution of rainfall through time.

sufficient latent heat energy to change from the liquid to the vapor phase and dissipate into the atmosphere. Transpiration is the release of water vapor into the atmosphere from plants. Soil moisture is drawn into plant roots by osmotic forces, moves from the roots through the plants to the leaves, and is expelled as vapor primarily through openings in the epidermis of leaves. Both evaporation and transpiration, that together are known as *evapotranspiration*, are drying processes that reduce soil moisture near the soil surface.

Evapotranspiration reduces antecedent soil moisture, thereby reducing runoff and erosion during rainfall events. Dry soil particles are much more susceptible than moist soil particles to detachment and entrainment by wind. Therefore, wind erosion is much less immediately following a rainstorm than during an extended dry period. The wetting and drying of the soil by rainfall, infiltration, and evapotranspiration increases the cohesion among soil particles following disturbance, thereby reducing soil erodibility.

SOURCES OF INFORMATION

There are various methods for computing the values of hydrologic variables used in soil-erosion analyses. Examples of these procedures, together with additional discussion, can be found in Dunne and Leopold (1978) and Haan et al. (1994), among other sources.

SUGGESTED READINGS

Dunne, T., and L. B. Leopold. 1978. *Water in Environmental Planning*. W. H. Freeman, San Francisco.

Haan, C. T., B. J. Barfield, and J. C. Hayes. 1994. *Design Hydrology and Sedimentology for Small Catchments*. Academic Press, San Diego, CA.

Linsley, R. K., M. A. Kohler, and J. L. H. Paulus. 1982. *Hydrology for Engineers*. McGraw-Hill, New York.

Maidment, D. R. (ed.). 1992. *Handbook of Hydrology*, McGraw-Hill, New York.

Skaggs, R. W., L. F. Huggins, E. J. Monke, and G. R. Foster. 1969. Experimental evaluation of infiltration equations. *Trans. ASAE* 12(6):822–838.

U.S. Department of Agriculture, Soil Conservation Service. 1972 (Revised 1985). *National Engineering Handbook, Section* 4, *Hydrology*. U.S. Government Printing Office, Washington, DC.

Appendix C: Soil Erosion Web Sites

Soil erosion research and soil conservation is under way throughout the world. The latest information concerning erosion research, measurement techniques, and the activities of erosion and related interest groups frequently can be found through Internet Web sites. Information posted on Web sites may predate publication by a year or more. In addition, there is the opportunity to contact persons actively engaged in research. The following is a list of Web site addresses, sponsoring organizations, and a brief comment concerning the Web site contents.

This list was compiled using addresses provided from colleagues responding to an open request on GEOMORPHLIST, our familiarity with various sites, and a Web search. Those receiving our request were encouraged to disseminate the call to others. We are grateful to those who responded. Undoubtedly, there are some important Web sites that escaped our attention. We apologize and will happily include these in the next edition of this book. All of the Web sites included in this list were active as of August 2001. The Web site addresses received were screened to remove those of commercial companies and individuals. It is not possible to include all of these, and it would be inappropriate to include only a few.

Sometimes, the comment indicates only that the site is the home page for the sponsoring organization. Home pages usually contain standard information concerning the mission, history, personnel, projects, and publications for the group. The project and publication directories often are quite extensive.

Web Site Address	*Sponsor*	*Comment*
www.agric.gov.ab.ca/agdex/500/72000002.html	Alberta Agriculture (Canada)	Introduction to wind-erosion control
www.ars.usda.gov	U.S. Department of Agriculture (USDA) Agricultural Research Service (ARS)	Home page
www.asce.org	American Society of Civil Engineers	Home page
www.blm.gov	U.S. Department of the Interior (USDI), Bureau of Land Management	Home page
www.brc.tamus.edu/swat/index.html	Grasslands, Soil and Water Research Laboratory, USDA, ARS, Temple, TX	Web site for SWAT (Soil and Water Assessment Tool)
www.ca.nrcs.usda.gov/wps/download.html	USDA, NRCS, California Unified Watershed Assessment	Clear water initiative, Clean Water Action Plan
www.ca.uky.edu/assmr	American Society for Surface Mining and Reclamation	Home page
www.cla.sc.edu/geog/gsgdocs	Geomorphology Specialty Group, Association of American Geographers	Home page
www.cost623.leeds.ac.uk/cost623	European Cooperation in the Field of Scientific and Technical Research (Cost Action 623)	Home page
www.csrl.ars.usda.gov/wewc	USDA, ARS, Wind Erosion and Water Conservation, Lubbock, TX	Home page
www.ctic.purdue.edu/CTIC/CTIC.html	Conservation Technology Information Center, nonprofit conservation organization	Home page
www.distromet.com	Distromet Ltd., Switzerland	Distributor of Joss–Waldvogel Rainfall impact disdrometer
www.distrometer.at	Joanneum Research CO, Austria	Distributor of ZD-Video distrometer
www.engr.utk.edu/research/water/erosion	University of Tennessee, Knoxville, Department of Water Resources Civil and Environmental Engineering	*Soil Erosion Prevention and Sediment Control* by J. L. Smoots and R. .D Smith (publication)

www.epa.gov	U.S. Environmental Protection Agency	Home page
www.erosioncontrol.com	International Erosion Control Association (IECA)	*Erosion Control*, Official Journal of the IECA
www.ex.ac.uk/~yszhang/caesium/welcome.htm	Geography Department at the University of Exeter, UK	Application of caesium-137 in soil erosion and sediment studies
www.fs.fed.us/rm	USDA, Forest Service, Rocky Mountain Research Station	Home page
www.ftw.nrcs.usda.gov/nhcp_2.html	USDA, NRCS	*National Handbook of Conservation Practices*, Conservation Practice Standards
www.ftw.nrcs.usda.gov/soils_data.html	USDA, NRCS	NRCS Data Resources: Soils
www.gcrio.org/geo/soil.html	U.S. Global Change Research Information Office	Soil and soil erosion, report and references
www.geog.le.ac.uk/bgrg	British Geomorphological Research Group	Rainfall simulation database
www.geog.ucl.ac.uk/weels	Wind Erosion on European Light Soils (WEELS)	Home page, soil sampling, monitoring mapping, measurement
www.geosociety.org	Geological Society of America	Home page
www.gov.on.ca/OMAFRA/english/engineer/soil/erosion.htm	Ministry of Agriculture, Food and Rural Affairs, Ottawa, Ontario, Canada	Engineering soil and water erosion, erosion control
www.homepage.montana.edu/~ueswl/geomorphlist/index.htm	International Association of Geomorphologists	Home page
www.ieca.org	International Erosion Control Association (IECA)	Home page
www.iscohome.org	International Soil Conservation Organization (ISCO)	Home page
www.iso.ch	International Organization for Standardization (ISO)	ISO 14000 Environmental Management Standards

Web Site Address	Sponsor	Comment
www.kuleuven.ac.be/facdep/geo/fgk/pages/expgeom.htm	Catholic University, Leuven, Belgium, Experimental Geomorphology Laboratory	Erosion research, modeling
www.lbk.ars.usda.gov/wewc	USDA, ARS, Wind Erosion and Water Conservation Lubbock, TX	Home page
www.lisa.univ-paris12.fr	Inter-university Laboratory of Atmospheric Systems, France	Home page (in French)
www.maf.govt.nz/MAFnet/schools/kits/soil.htm	Ministry of Agriculture, New Zealand	Home page, soil erosion
www.medalus.demon.co.uk	Mediterranean Desertification and Land Use Project	Home page
www.mrsars.usda.gov	USDA, ARS, North Central Soil Conservation Research Laboratory, Morris, MN	Home page
www.nal.usda.gov/wqic/wqdb/esearch.html	USDA, ARS Water Quality Information Center National Agricultural Library, Beltsville, MD	Database of online documents covering water and agriculture
www.ncg.nrcs.usda.gov	USDA, NRCS Fort Worth, TX	Home page
www.ncg.nrcs.usda.gov/tech_tools.html	USDA, NRCS	Technical tools; download files
www.nrcs.usda.gov	USDA, NRCS	Home page
www.nrcs.usda.gov/TechRes.html	USDA, NRCS	Technical Resources, National Science and Technology Consortium
www.odyssey.maine.edu/gisweb/spatdb/egis/eg94023.html	Utrecht University, Department of Geography, Utrecht, The Netherlands	LISEM: physically based hydrological and soil erosion model incorporated in a GIS
www.osei.noaa.gov/Events/Dust	U.S. Department of Commerce National Oceanic and Atmospheric Administration	Dust storms, satellite Imagery
www.osmre.gov	USDI, Office of Surface Mining	Home page

www.psw.fs.fed.us	USDA, Forest Service, Pacific Southwest Research Station	Home page
www.psw.fs.fed.us/techpub.html	USDA, Forest Service, Pacific Southwest Research Station	Publication list
www.rusle2.com	University of Tennessee, Knoxville Agricultural and Biosystems Engineering Department	RUSLE 2 soil loss program
www.sedlab.olemiss.edu	USDA, ARS National Sedimentation Laboratory Oxford, MS	Home page, Revised Universal Soil Loss Equation, Agricultural Non-Point-Source Pollution Model (AGNPS)
www.sedlab.olemiss.edu/AGNPS2001/concepts/concepts.html	USDA, ARS National Sedimentation Laboratory Oxford, MS	Conservational Channel Evolution and Pollutant Transport System (Concept), Agricultural Non-Point-Source Pollution Model (AGNPS)
www.sedlab.olemiss.edu/cwp.html	USDA, ARS National Sedimentation Laboratory, Oxford, MS	Channel and watershed process
www.sedlab.olemiss.edu/rusle	USDA, ARS, National Sedimentation Laboratory, Oxford, MS	Revised Universal Soil-Loss Equation (RUSLE)
www.shef.ac.uk/~scidr	University of Sheffield, Sheffield Center for International Drylands Research, UK	Home page
www.silsoe.cranfield.ac.uk/iwe/erosion/eurosem.htm	Institute of Water and Environment, Cranfield University, Silsoe, UK	Eurosem erosion model, Home page
www.soilerosion.net	Queen's University of Belfast–Northern Ireland, School of Geography	Home page (news, photos)
www.statlab.iastate.edu/soils/cer	Spatial Frameworth of Ecological Units for the United States National Interagency Commit-	Ecological regions of conterminous United States

Web Site Address	Sponsor	Comment
www.stream.fs.fed.us	USDA, Forest Service, Stream Team	Links to other sites; hydrology, geomorphology
www.swcs.org	Soil and Water Conservation Society	Home page
www.tucson.ars.ag.gov	USDA, ARS Southwest Watershed Research Center, Tucson, AZ	Home page
www.usace.army.mil	U.S. Army Corps of Engineers	Home page
www.usbr.gov	USDI, U.S. Bureau of Reclamation	Home page
www.usgs.gov	USDI. U.S. Geological Survey, Reston, VA	Home page
www.watershed.org/wmc	Watershed Management Council	Watershed management streams
www.wcc.nrcs.usda.gov/water/climate/gem/gem.html	USDA, NRCS, National Water and Climate Center	Weather generator, GEM
www.wcc.nrcs.usda.gov/water/quality/frame/wqam	USDA, NRCS, National Water and Climate Center, Portland, OR	Home page: water quality assessment and monitoring
www.wcc.nrcs.usda.gov/wtec/wtec.html	USDA, ARS	Watershed technology electronic catalog, practices
www.weru.ksu.edu/nrcs	USDA, NRCS, Wind Erosion Research Unit Manhattan, KS	NRCS-ARS wind erosion information exchange site
http://asae.org	American Society of Agricultural Engineers	Home page
http://boris.qub.ac.uk/bgrg	British Geomorphological Research Group (BGRG)	Home page
http://cmex-www.arc.nasa.gov/Aeolian/Aeolian.html	Arizona State University Planetary Aeolian Laboratory	Home page
http://dillaha.bse.vt.edu/answers/index.htm	Biological Systems Engineering Department Virginia Technological University	ANSWERS model Home page
http://dmoz.org/Science/Earth_Sciences/Geology/Geomorphology/Soil_Erosion	Dmoz Open Directory Project Geomorphology, Soil Erosion	Links to other site

http://forest.moscowfsl.wsu.edu/4702/wepp0.html	USDA, Forest Service, Rocky Mountain Research Station Forestry Sciences Laboratory Moscow, ID	Home page, erosion modeling Water Erosion Prediction Program (WEPP)
http://grl.ars.usda.gov	USDA, ARS, Grazing Lands Research Laboratory Ft. Reno, OK	Home page
http://hydrolab.arsusda.gov	USDA, ARS, Hydrology and Remote Sensing Laboratory Beltsville, MD	Home page: hydrology, remote sensing
http://hydrolab.arsusda.gov/wdc/arswater.html	USDA, ARS, Water Data Center Beltsville, MD	ARS water database
http://kimberly.ars.usda.gov/pampage.shtml	USDA, ARS, Northwest Irrigation and Soils Research Laboratory, Kimberly, ID	PAM research project
http://mwnta.nmw.ac.uk/GCTEFocus3/networks/erosion.htm	Global Change and Terrestial Ecosystems Focus 3	Home page: soil erosion network
http://office.geog.uvic.ca/dept/cgrg/cgrg.htm	Canadian Geomorphology Research Group, Ontario–Canada	Home page
http://pubs.usgs.gov/gip/deserts/eolian	USDI, U.S. Geological Survey	Eolian processes
http://topsoil.nserl.purdue.edu	USDA, ARS, National Soil Erosion Research Laboratory	Home page
http://topsoil.nserl.purdue.edu/gis	USDA, ARS, National Soil Erosion Research Laboratory	GeoWepp, Geo-Spatial Interface for Water Erosion Prediction Program (WEPP)
http://topsoil.nserl.purdue.edu/iscohome/index.html	USDA, ARS, National Soil Erosion Research Laboratory	International Soil Conservation Organization Home page
http://tucson.ars.ag.gov/kineros	USDA, ARS	Kineros model home page
http://water.usgs.gov/osw	USDI, U.S. Geological Survey	Surface water information
http://weru.ksu.edu	USDA, ARS, Wind Erosion, Research Unit, Manhattan, KS	Home page

References

Agassi, M. (ed.). 1996. *Soil Erosion, Conservation, and Rehabilitation*. Marcel Dekker, New York.

Agnew, B. 2000. *Practical Approaches for Effective Erosion and Sediment Control*. International Erosion Control Association, Steamboat Springs, CO.

Alberts, E. E., W. C. Moldenhauer, and G. R. Foster. 1980. Soil aggregates and primary particles transported in rill and interrill erosion. *Soil Sci. Soc. Am. J.* 44:590–595.

Al-Durrah, M. M., and J. M. Bradford. 1982. The mechanism of raindrop splash on soil surfaces. *Soil Sci. Soc. Am. J.* 46:1086–1090.

Anhert, F. 1988. Modeling landform change. In: M. G. Anderson (ed.), *Modeling Geomorphological Systems*, pp. 375–400, Wiley, Chichester, West Sussey England.

Armbrust, D. V., and L. Lyles. 1975. Soil stabilizers to control wind erosion. *Soil Cond.* 7:77–82.

Arndt, C., B. Fecso, P. V. Preckel, and B. Stoneman. 2001. Soil selection for use in environmental analysis. *J. Soil Water Conserv.* 56:165–170.

Arnold, J. G., R. Shrinivasan, R. S. Muttiah, and J. R. Williams. 1998. Large area hydrologic modeling and assessment. I. Model development. *J. Am. Water Resour. Assoc.* 34:73–89

Austin, M. E. 1981. *Land Resource Regions and Major Land Resource Areas of the United States*. USDA Agricultural Handbook 296. U.S. Government Printing Office, Washington, DC.

Barbier, E. B., and J. T. Bishop. 1995. Economic values and incentives affecting soil and water conservation in developing countries. *J. Soil Water Conserv.* 50(2):133–137.

Barker, R., L. Dixon, and J. Hooke. 1997. Use of terrestrial photogrammetry for monitoring and measuring bank erosion. *Earth Surf. Processes Landforms* 22: 1217–1227.

Bauer, B. O., and S. L. Namikas. 1998. Design and field test of a continuously weighing, tipping-bucket assembly for aeolian sand traps. *Earth Surf. Processes Landforms* 23:1171–1183.

Bennett, H. H., and W. C. Lowdermilk. 1938. General aspects of the soil-erosion problem. *Soils and Men: 1938 Yearbook of Agriculture*. U.S. Government Printing Office, Washington, DC.

Bilbro, J. D., and D. W. Fryrear. 1994. Wind erosion losses are related to plant silhouette and soil cover. *Agron. J.* 86: 550–553.

Bingner, R. L., L. J. Garbrecht, J. G. Arnold, and R. Shrinivasan. 1997. Effects of watershed subdivision on simulated runoff and fine sediment yield. *Trans. ASAE* 40:1329–1335.

Birkeland, P. W. 1999. *Soils and Geomorphology*, 3rd ed. Oxford University Press, New York.

Boardman, J. 1998. An average soil erosion rate for Europe: myth or reality. *J. Soil Water Conserv.* 53:46–50.

Bollinne, A. 1980. Splash measurements in the field. In: M. De Boodt and D. Gabriels (eds.), *Assessment of Erosion*, pp. 441–453. Wiley, New York.

Bosch, D., F. Theurer, R. Bingner, G. Felton, and I. Chaubey. 1998. *Evaluation of the AnnAGNPS Water Quality Model*. Paper 98–2195. American Society of Agricultural Engineers, St. Joseph, MI.

Bowie, A. J. 1995. *Use of Vegetation to Stabilize Eroding Streambanks*. USDA Conservation Research Report 43. U.S. Government Printing Office, Washington, DC.

Box, J. E. 1981. The effect of surface slaty fragments on soil erosion by water. *Soil Sci. Soc. Am. J.* 45:111–116.

Bradford, J. M., and C. Huang. 1992. Mechanisms of crust formation: physical components. In: M.D. Sumner and B. A. Stewart (eds.), *Soil Crusting: Chemical and Physical Processes*, pp. 55–72. Lewis Publishers, Boca Raton, FL.

Bradford, J. M., and C. Huang. 1993. Comparison of interrill soil loss for laboratory and field procedures. *Soil Technol.* 6:145–156.

Bradford, J. M., P. A. Remley, J. E. Ferris, and J. F. Santini. 1986. Effects of soil surface sealing on splash from a single waterdrop. *Soil Sci. Soc. Am. J.* 50:1547–1552.

Brakensiek, D. L., H. B. Osborn, and W. J. Rawls (coordinators). 1979. *Field Manual for Research in Agricultural Hydrology*. USDA Agricultural Handbook 224. U.S. Government Printing Office, Washington, DC.

Brown, L. C., and G. R. Foster. 1987. Storm erosivity using idealized intensity distributions. *Trans. ASAE* 30(2):379–386.

Brown, L. C. and H. Kane. 1994. *Full House: Reassessing the Earth's Population Carrying Capacity*. World Watch Institute, Washington, DC.

Brown, L. C., G. R. Foster, and D. B. Beasley. 1989. Rill erosion as affected by incorporated crop residue and seasonal consolidation. *Trans. ASAE* 32(6):1967–1978.

Bryan, R. B. 1977. Assessment of soil erodibility: new approaches and directions. In: T. J. Toy (ed.), *Erosion: Research Techniques, Erodibility and Sed-*

iment Delivery, pp. 57–72. Geo Books, imprint of Geo Abstracts Ltd., *Norwich, England.*

Bubenzer, G. D. 1979a. Inventory of rainfall simulators. In: *Proc. Rainfall Simulator Workshop*, Mar. 7–9, 1979, pp. 120–130. USDA, Agricultural Research Service, Agricultural Reviews and Manuals, ARM-W-10/July 1979. U.S. Government Printing Office, Washington, DC.

Bubenzer, G. D. 1979b. Rainfall characteristics important for simulation. In: *Proc. Rainfall Simulator Workshop*, Mar. 7–9, pp. 22–34. USDA Agricultural Research Service, Agricultural Reviews and Manuals, ARM-W-10/July 1979. U.S. Government. Printing Office, Washington, DC.

Campbell, I. A. 1970. Micro-relief measurements on unvegetated shale slopes. *Prof. Geogr.* 22(4):215–220.

Campbell. I. A. 1974. Measurements of erosion on badland surfaces. *Z. Geomorphol.*, Suppl. Bd. 21:122–137.

Chapman, G. 1948. Size of raindrops and their striking force at the soil surface in a red pine plantation. *Trans. Am. Geophys. Union* 29:664–670.

Chepil, W. S. 1956. Influence of moisture on erodibility of soil by wind. *Soil Sci. Soc. Am. Proc.* 20:288–292.

Chien, N., and Z. Wan. 1999. *Mechanics of Sediment Transport*. ASCE Press. Reston, VA.

Chorley, R. J. 1966. The application of statistical methods to geomorphology. In: G. H. Drury (ed.), *Essays in Geomorphology*, pp. 275–388. Heinemann, London.

Chow, V. T. 1959. *Open-Channel Hydraulics*. McGraw-Hill, New York.

Colorado Department of Highways. 1978. *Erosion Control Manual*. CDH, Denver, CO.

Commission on Long-Range Soil and Water Conservation, Board on Agriculture, National Research Council. 1993. *Soil and Water Quality: An Agenda for Agriculture*. National Academy Press. Washington, DC.

Commission on Watershed Management, National Research Council. 1999. *New Strategies for America's Watersheds*. National Academy Press, Washington, DC.

Cook, M. J., and W. H. Valentine. 1979. Monitoring cut slope erosion by close range photogrammetry. *Field Notes U.S. For. Serv.* 11(7):5–9.

Crosson, P. 1995. Soil erosion estimates and cost. *Science* 269: 461–464.

Curtis, W. R. 1971. Strip-mining, erosion, and sedimentation. *Trans. ASAE* 14(3):434–436.

Curtis W. R., K. L. Dyer, and G. P. Williams. Undated. *A Manual for Training Reclamation Inspectors in the Fundamentals of Hydrology*. U.S. Department of Agriculture, Forest Service. Prepared for Office of Surface Mining Enforcement and Reclamation. U.S. Government Printing Office, Washington, DC.

Dabney, S. M., K. C. McGregor, L. D. Meyer, E. H. Grissinger, and G. R. Foster. 1993. Vegetative barriers for runoff and sediment control. In: *Proc. International Symposium, Integrated Resource Management*, pp. 60–70. American Society of Agricultural Engineers, St. Joseph, MI.

Davis, S. S., G. R. Foster, and L. F. Huggins. 1983. Deposition of nonuniform sediment on concave slopes. *Trans. ASAE* 26(4):1057–1063.

DeBano, L. F. 1981. *Water Repellent Soils: A State-of-the-Art*. USDA Forest Service General Technical Report PSW-46. Pacific Southwest Forest and Range Experiment Station, Berkeley, CA.

Dillaha, T. A., B. B. Ross, S. Mostaghimi, C. D. Heatwole, and V. O. Shanholtz. 1988. Rainfall simulation: a tool for best management practice education. *J. Soil Water Conserv*. 43(4):288–290.

Dissmeyer, G. E., and G. R. Foster. 1980. *A Guide for Predicting Sheet and Rill Erosion on Forest Land*. USDA Forest Service Technical Publication SA-TP-11. U.S. Government Printing Office, Washington, DC.

Dissmeyer, G. E., and G. R. Foster. 1981. Estimating the cover-management factor (*C*) in the Universal Soil Loss Equation for forest conditions. *J. Soil Water Conserv*. 36:235–240.

Dudal, R. 1981. An evaluation of conservation needs. In: R. P. C. Morgan (ed.), *Soil Conservation: Problems and Prospects*. Wiley, New York.

Dunne, T. 1998. Critical data requirements for prediction of erosion and sedimentation in mountain drainage basins. *J. Am. Water Resour. Assoc*. 34: 795–808.

Dunne, T., and L. B. Leopold. 1978. *Water in Environmental Planning*. W. H. Freeman, San Francisco.

Edwards, W. M., and L. B. Owens. 1991. Large storm effects on total soil erosion. *J. Soil Water Conserv*. 46(1):75–78.

Eisenhart, C. 1952. The reliability of measured values. I. Fundamental concepts. *Photogramm. Eng*. 18:542–554.

Elliot, W. J. 1999. Understanding and modeling erosion from insloping roads. *J. For*. 97:30–34.

El Swaify, S. A., and D. S. Yakowitz (eds.). 1998. *Multiple Objective Decision Making for Land, Water, and Environmental Management*. Davis Publishers, New York.

El-Swaify, S. A., E. W. Dangler, and C. L. Armstrong. 1982. *Soil Erosion by Water in the Tropics*. College of Tropical Agriculture and Human Resources, University of Hawaii, Honolulu, HI.

El-Swaify, S. A., W. C. Moldenhauer, and A. Lo. 1983. *Soil Erosion and Conservation*. Soil and Water Conservation Society Ankeny, IA.

Emmett, W. W. 1970. *The Hydraulics of Overland Flow on Hillslopes*. USGS Professional Paper 662A. U.S. Government Printing Office, Washington, DC.

Esseks, J. D., S. E. Kraft, and D. M. Ihrke. 2000. Policy lessons from a quasi-regulatory conservation. In: T. L. Napier, S. M. Napier, and J. Tvrdon (eds.), *Soil and Water Conservation Policies and Programs: Successes and Failures*, pp. 109–125. Soil and Water Conservation Society, Ankeny, IA.

Flanagan, D. C., G. R. Foster, and W. C. Moldenhauer. 1988. Storm pattern effect on infiltration, runoff, and erosion. *Trans. ASAE* 31(2):414–420.

Flanagan, D. C., G. R. Foster, W. H. Neibling, and J. P. Burt. 1989. Simplified equations for filter strip design. *Trans. ASAE* 32(6):2001–2007.

Follett, R. F., and B. A. Stewart (eds.). 1985. *Soil Erosion and Crop Productivity*. American Society of Agronomy, Crop Science Society of America, Soil Science of Society of America, Madison, WI.

Foster, G. R. 1971. The overland flow processes under natural conditions. In: *Biological Effects in the Hydrological Cycle: Proc. 3rd International Seminar for Hydrological Professors*, pp. 173–185. Purdue University, West Lafayette, IN.

Foster, G. R. 1980. Soil erosion modeling: special considerations for nonpoint pollution evaluation on field sized areas. In: M. R. Overcash and J. M. Davidson (eds.), *Environmental Impact of Nonpoint Source Pollution*, pp. 213–240. Ann Arbor Science, Ann Arbor, MI.

Foster, G. R. 1982. Modeling the erosion process. In: C. T. Haan, H. P. Johnson, and D. L. Brakensiek (eds.), *Hydrologic Modeling of Small Watersheds*, pp. 297–382. American Society of Agricultural Engineers, St. Joseph, MI.

Foster, G. R. 1985. Understanding ephemeral gully erosion (concentrated flow erosion). In: *Soil Conservation: Assessing the National Resources Inventory*, pp. 90–125. National Academy Press, Washington, DC.

Foster, G. R., and R. E. Highfill. 1983. Effect of terraces on soil loss: USLE *P*-factors for terraces. *J. Soil Water Conserv*. 38:48–51.

Foster, G. R., and L. D. Meyer. 1975. Mathematical simulation of upland erosion by fundamental erosion mechanics. In: *Present and Prospective Technology for Predicting Sediment Yields and Sources*, pp. 190–204. ARS-S-40. USDA Science and Education Administration, Washington, DC.

Foster, G. R., L. D. Meyer, and C. A. Onstad. 1977. An erosion equation derived from basic erosion principles. *Trans. ASAE* 20(4):678–682.

Foster, G. R., F. P. Eppert, and L. D. Meyer. 1979. A programmable rainfall simulator for field plots. In: *Proc. Rainfall Simulator Workshop*, Mar. 7–9, 1979, pp. 45–59. USDA Agricultural Research Service, Agricultural Reviews and Manuals, ARM-W-10/July 1979. U.S. Government Printing Office, Washington, DC.

Foster, G. R., W. H. Neibling, S. S. Davis, and E. E. Alberts. 1980. Modeling particle segregation during deposition by overland flow. In: *Proc. Hydrologic Transport Modeling Symposium*, pp. 184–195. American Society of Agricultural Engineers, St. Joseph, MI.

Foster, G. R., F. Lombardi, and W. C. Moldenhauer. 1982. Evaluation of rainfall-runoff erosivity factors for individual storms. *Trans ASAE* 25(1): 124–129.

Foster, G. R., L. F. Huggins, and L. D. Meyer. 1984. A laboratory study of rill hydraulics. II. Shear stress relationships. *Trans. ASAE* 37(3):797–804.

Foster, G. R., R. A. Young, and W. H. Neibling. 1985a. Sediment composition for nonpoint source pollution analyses. *Trans. ASAE* 28(1):133–139, 146.

Foster, G. R., G. C. White, T. E. Hakonson, and M. Dreicer. 1985b. A model for splash and retention of sediment and soil-borne contaminants on plants. *Trans. ASAE* 28(5):1511–1520.

Foster, G. R., R. A. Young, M. J. M. Römkens, and C. A. Onstad. 1985c. Processes of soil erosion by water. In: R. F. Follett and B. A. Stewart, (eds.), *Soil Erosion and Crop Productivity*, pp. 137–162. American Society of Agronomy, Crop Science Society of America, Soil Science Society of America, Madison, WI.

Foster, G. R., G. A. Weesies, K. G. Renard, D. C. Yoder, D. K. McCool, and J. P. Porter. 1997. Support practices. In: *Predicting Soil Erosion by Water: A Guide to Conservation Planning with the Revised Universal Soil Loss Equation (RUSLE)*, pp. 183–251. USDA Agricultural Handbook 703. U.S. Government Printing Office, Washington, DC.

Foth, H. D., and L. M. Turk. 1972. *Fundamentals of Soil Science*. Wiley, New York.

Franti, T. G., G. R. Foster, and E. J. Monke. 1996. Modeling the effects of incorporated residue on rill erosion. II. Experimental results and model validation. *Trans. ASAE* 39:543–550.

Free, G. R. 1960. Erosion characteristics of rainfall. *Agric. Eng.* 41:447–449, 455.

Fryrear, D. W., and E. L. Skidmore. 1985. Methods for controlling wind erosion. In: R. F. Follett and B. A. Stewart (eds.), *Soil Erosion and Crop Productivity*, pp. 443–457. American Society of Agronomy, Crop Science Society of America, Soil Science Society of America, Madison, WI.

Fryrear, D. W., J. E. Stout, L. J. Hagen, and E. D. Vories. 1990. Wind erosion: field measurement and analysis. *Trans. ASAE* 3:155–160.

Gibbens, R. P., J. M. Tromble, J. T. Hennesy, and M. Cardenas. 1983. Soil movement in mesquite dunelands and former grasslands of Southern New Mexico from 1933 to 1980. *J. Range Manage*: 36(2):145–148.

Govers, G., and J. Poesen. 1986. A field-scale study of surface sealing and compaction of loam and sandy loam soils. I. Spatial variability of surface sealing and crusting. In: *Assessment of Soil Surface Sealing and Crusting, Proc. Symposium*, pp. 171–182. Flanders Research Centre for Soil Erosion and Soil Conservation, Gent, Belgium.

Grace, J. M., III. 2000. Forest road sideslopes and soil conservation techniques. *J. Soil Water Conserv*. 55:96–101.

Gregory, S. 1971. *Statistical Methods and the Geographer*, 2nd ed. Longman Group, London.

Griffin, M. L., D. B. Beasley, J. J. Fletcher, and G. R. Foster. 1988. Estimating soil loss for topographically nonuniform field and farm units. *J. Soil Water Conserv*. 43(4):326–331.

Grissenger, E. H. 1966. Resistance of selected clay systems to erosion by water. *Water Resour. Res. Am. Geophys. Union* 2:131–138.

Haan, C. T., B. J. Barfield, and J. C. Hayes. 1994. *Design Hydrology and Sedimentology for Small Catchments*. Academic Press, San Diego, CA.

Haering, K. C., W. L. Daniels, and S. E. Feagley. 2000. Reclaiming mined lands with biosolids, manures, and papermill sludge. In: R. I. Barnhisel, R. G. Darmody, and W. L. Daniels (eds.), *Reclamation of Drastically Disturbed Lands*, pp. 615–644. American Society of Agronomy, Crop Science Society of America, Soil Science Society of America, Madison WI.

Hagen, L. J. 1976. Windbreak design for optimum wind erosion control. In: *Shelterbelts on the Great Plains: Proc. Symposium*, pp. 31–36. Publication 78. Great Plains Agricultural Council, Lincoln, NE.

Hagen, L. J., E. L. Skidmore, and A. Saleh. 1992. Prediction of aggregate abrasion coefficients. *Trans. ASAE* 35:1847–1850.

Haigh, M. J. 1977. *The Use of Erosion Pins in the Study of Slope Evolution*. British Geomorphological Research Group Technical Bulletin 18, pp. 31–49. Geo Abstracts Ltd., Norwich, England.

Hairsine, P. B. and C. W. Rose. 1991. Rainfall detachment and deposition: sediment transport in the absence of flow-driven processes. *Soil Sci. Soc. Am. J*. 55:320–324.

Hall, G. F., T. J. Logan, and K. K. Young. 1985. Criteria for determining tolerable erosion rates. In: R. F. Follett and B. A. Stewart (eds.), *Soil Erosion and Productivity*, pp. 173–188. American Society of Agronomy, Crop Science Society of America, Soil Science Society of America, Madison, WI.

Heede, B. H. 1975. Stages of development of gullies in the west. In: *Present and Prospective Technology for Predicting Sediment Yields and Sources*, pp. 155–161. USDA Agricultural Research Service ARS-S-40. USDA Science and Education Administration, Washington, DC.

Heimlich, R. E. 1985. Soil erosion on new cropland: a sodbusting perspective. *J. Soil Water Conserv*. 40(4):322–326.

Hennessy, J. T., B. Kies, R. P. Gibbens, and J. M. Tromble. 1986. Soil sorting by forty-five years of wind erosion on a southern New Mexico range. *Soil Sci. Soc. Am. J*. 50(2):391–394.

Herbel, C. H., and W. L. Gould. 1995. *Management of Mesquite, Creosotebush, and Tarbush with Herbicides in the Northern Chihuahuan Desert*. Agricultural Experiment Station Bulletin 775. New Mexico State University, Las Cruces, NM.

Heusch, B. 1981. Sociological constraints in soil conservation: a case study, the Rif Mountains, Morocco. In: R. P. C. Morgan (ed.), *Soil Conservation: Problems and Prospects*, pp. 419–424. Wiley, New York.

Hjelmfelt, A. T., L. A. Kramer, and R. G. Spomer. 1986. Role of large events in average soil loss. In: *Proc. 4th Federal Interagency Sedimentation Conference*, Las Vegas, NV, Vol. 1, pp. 3–1 to 3–10. Interagency Advisory Committee on Water Data.

Horton, R. E. 1933. The role of infiltration in the hydrologic cycle. *Trans. Am. Geophys. Union* 20: 693–711.

Huang, C., and J. M. Bradford. 1990. Portable laser scanner for measuring soil surface roughness. *Soil Sci. Soc. Am. J.* 54:1402–1406.

Huang, C., and J. M. Laflen. 1996. Seepage and soil erosion for a clay loam soil. *Soil Sci. Soc. Am. J.* 60:408–416.

Hudson, N. W. 1965. *Field Measurements of Accelerated Soil Erosion in Localized Areas*. Rhodesia Southern Ministry of Agricultural Research Bulletin 2249.

Hudson, N. W. 1981a. *Instrumentation for Studies of the Erosive Power of Rainfall*. Bulletin 133, International Association of Hydrological Sciences, pp. 383–390. Washington, DC.

Hudson, N. W. 1981b. Social, political and economic aspects of soil conservation. In: R. P. C. Morgan (ed.), *Soil Conservation: Problems and Prospects*, pp. 45–54. Wiley, New York.

Hudson, N. W. 1993. *Field Measurement of Soil Erosion and Runoff*. Food and Agriculture Organization of the United Nations, Rome.

Hurni, H. 2000. Soil conservation policies and sustainable land management: A global overview. In: T. L. Napier, S. M. Napier, and J. Tvrdon (eds.), *Soil and Water Conservation Policies and Programs: Successes and Failures*, pp. 19–30. Soil and Water Conservation Society Ankeny, IA.

Johnson, L. C. 1987. Soil loss tolerance: fact or myth? *J. Soil Water Conserv.* 41:336–338.

Johnson, G. L., C. L. Hanson, S. P. Hardegree, and E. B. Ballard. 1996. Stochastic weather simulation: overview and analysis of two commonly used models. *J. Appl. Meteorol.* 35:1878–1896.

Kebede, T., and H. Hurni (eds.). 1992. *Soil Conservation for Survival*. Soil and Water Conservation Society, Ankeny, IA.

Kemper, W. D., R. Roseman, and S. Nelson. 1985. *Gas displacement and Aggregate Stability of Soils*. Soil Science Soc. J. 49:25–28.

Khan, M. J., E. J. Monke, and G. R. Foster. 1988. Mulch cover and canopy effect on soil loss. *Trans. ASAE* 31:706–711.

Kincaid, D. C., K. H. Solomon, and J. C. Oliphant. 1996. Drop size distributions for irrigation sprinklers. *Trans. ASAE* 39(3):839–845.

Knighton, D. 1984. *Fluvial Forms and Processes*. Edward Arnold, London.

Knisel, W. G., K. G. Renard, and L. J. Lane. 1979. Hydrologic data collection: how long is long enough? In: *Specialty Conference, Irrigation and Drainage in the Nineteen-Eighties*, pp. 238–254. American Society of Civil Engineers, Albuquerque, NM.

Knott, P., and A. Warren. 1981. Aeolian processes. In: A. Goudie (ed.), *Geomorphological Techniques*, pp. 226–246. George Allen & Unwin, London.

Kral, D. M. (ed.). 1982. *Determinants of Soil Loss Tolerance*. ASA Special Publication 45. American Society of Agronomy, Madison, WI.

Kuhnle, R. A., R. L. Bingner, G. R. Foster, and E. H. Grissenger. 1996. Effect of land use changes on sediment transport in Goodwin Creek. *Water Resour. Res.* 32(10):3189–3196.

Laflen, J. M., W. C. Moldenhauer, and T. S. Colvin. 1980. Conservation tillage and soil erosion on continuously row-cropped land. In: *Crop Production with Conservation in the 80's*, pp. 121–133. American Society of Agricultural Engineers, St. Joseph, MI.

Laflen, J. M., L. J. Lane, and G. R. Foster. 1991a. WEPP: a new generation of erosion prediction technology. *J. Soil Water Conserv.* 46(1):34–38.

Laflen, J. M., W. J. Elliot, J. R. Simanton, C. S. Holzhey, and D. Kohl. 1991b. WEPP soil erodibility experiments for rangeland and cropland soils. *J. Soil Water Conserv.* 46(1):39–44.

Lal, R. 1985. Soil erosion in its relationship to productivity in tropical soils. In: S. A. El Swaify, W. C. Moldenhauer, and A. Lo (eds.), *Soil Erosion and Conservation*, pp. 237–247. Soil and Water Conservation Society, Ankeny, IA.

Lal, R. 1994a. Soil erosion by wind and water: problems and prospects. In: R. Lal, (ed.), *Soil Erosion Research Methods*, 2nd ed. Soil and Water Conservation Society and St. Lucie Press, Ankeny, IA.

Lal, R. (ed.). 1994b. *Soil Erosion Research Methods*, 2nd ed. Soil and Water Conservation Society and St. Lucie Press, Ankeny, IA.

Lal, R., and F. J. Pierce. 1991. *Soil Management for Sustainability*. Soil and Water Conservation Society, Ankeny, IA.

Larson, W. E., G. R. Foster, R. R. Allmaras, and C. M. Smith (eds.). 1990. *Proc. Soil Erosion and Productivity Workshop*. University of Minnesota, St. Paul, MN.

Larson, W. E., M. J. Lindstrom, and T. E. Schumacher. 1997. The role of severe storm in soil erosion: a problem needing consideration. *J. Soil Water Conserv.* 52:90–95.

Lattanzi, A. R., L. D. Meyer, and M. F. Baumgardner. 1974. Influence of mulch rate and slope steepness on interrill erosion. *Soil Sci. Soc. Am. Proc.* 36:846–850.

Laws, J. O. 1941. Measurement of fall velocity of water drops and raindrops. *Trans. Am. Geophys. Union* 22:709–721.

Lindstrom, M. J., W. W. Nelson, and T. E. Schmacher. 1992. Quantifying tillage erosion rates due to moldboard plowing. *Soil Tillage Res.* 24:243–255.

Linsley, R. K., M. A. Kohler, and J. L. H. Paulhus. 1982. *Hydrology for Engineers*. McGraw-Hill, New York.

Little, W. C. and J. B. Murphey. 1982. Model study of low drop grade control structures. *Proc. ASCE, J. Hydraul Div.* 108(HY10):1132–1146.

Lu, J. Y., W. H. Neibling, G. R. Foster, and E. A. Cassol. 1988. Selective transport and deposition of sediment particles by shallow flow. *Trans. ASAE* 31(4):1141–1147.

Lusby, G. C., and T. J. Toy. 1976. An evaluation of surface-mine spoils area restoration in Wyoming using rainfall simulation. *Earth Surf. Processes Landforms* 1:375–386.

Lyles, L., and J. Tatarko. 1986. Wind erosion effects on soil texture and organic matter. *J. Soil Water Conserv.* 41:191–193.

Lyles, L., D. V. Armbrust, J. D. Dickerson, and N. P. Woodruff. 1969. Spray-on adhesives for temporary wind erosion control. *J. Soil Water Conserv.* 24: 190–193.

Lyles, L., L. J. Hagen, and E. L. Skidmore. 1983. Soil conservation: principles of erosion by wind. In: *Dryland Agriculture*. Agronomy Monograph 23, pp. 177–188. American Society of Agronomy, Madison, WI.

Lyles, L., G. W. Cole, and L. J. Hagen. 1985. Wind erosion: processes and prediction. In: R. F. Follett and B. A. Stewart (eds.), *Soil Erosion and Crop Productivity*, pp. 163–172. American Society of Agronomy, Crop Science Society of America, Soil Science Society of America, Madison, WI.

Maidment, D. R. (ed.). 1992. *Handbook of Hydrology*. McGraw-Hill, New York.

Mannering, J. V. 1981. The use of soil loss tolerance as a strategy for soil conservation. In: R. P. C. Morgan (ed.), *Soil Conservation: Problems and Prospects*, pp. 337–349. Wiley, New York.

McCool, D. K., M. G. Dossett, and S. J. Yecha. 1981. A portable rill meter for measuring field soil loss. In: *Erosion and Sediment Transport Measurement: Proc. Florence Symposium*. Publication 133, pp. 479–484. International Association of Hydrological Sciences, Florence, Italy.

McCool, D. K., L. C. Brown, G. R. Foster, C. K. Mutchler, and L. D. Meyer. 1987. Revised slope steepness factor for the Universal Soil Loss Equation. *Trans. ASAE* 30(4):1387–1396.

McCool, D. K., G. R. Foster, C. K. Mutchler, and L. D. Meyer. 1989. Revised slope length factor for the Universal Soil Loss Equation. *Trans. ASAE* 32(5): 1571–1576.

McCool, D. K., M. T. Walter, and L. G. King. 1995. Runoff index values for frozen soil areas of the Pacific Northwest. *J. Soil Water Conserv*. 50:466–469.

McCool, D. K., K. E. Saxton, and J. D. Williams. 1997. Surface cover effects on soil loss from temporally frozen cropland in the Pacific Northwest. In: *Proc. International Symposium on Physics, Chemistry, and Ecology of Seasonally Frozen Soils*, pp. 235–341. Cold Regions Research and Engineering Laboratory, U.S. Department of the Army. Hanover, NH.

McCormack, D. E., and K. K. Young. 1981. Technical and societal implications of soil loss tolerance. In: R. P. C. Morgan (ed.), *Soil Conservation: Problems and Prospects*, pp. 364–376. Wiley, New York.

McGregor, K. C., R. L. Bengston, and C. K. Mutchler. 1990. Surface and incorporated wheat straw effect on interrill runoff and soil erosion. *Trans. ASAE* 33:469–474.

Mead, R. H., and R. S. Parker. 1985. Sediment in rivers of the United States. In: *National Water Summary, 1984: Water Quality Issues*. USGS Water Supply Paper 2275, pp. 49–60. US. Government Printing Office, Washington, DC.

Meyer, L. D. 1981. How intensity affects interrill erosion. *Trans. ASAE* 25: 1472–1475.

Meyer, L. D. 1994. Rainfall simulators for soil erosion research. In: R. Lal (ed.), *Soil Erosion Research Methods*, 2nd ed., pp. 83–103. Soil and Water Conservation Society and St. Lucie Press, Ankeny, IA.

Meyer, L. D., and W. C. Harmon. 1985. Sediment losses from cropland furrows of different gradients. *Trans. ASAE* 28:448–453, 461.

Meyer, L. D., and D. L. McCune. 1958. Rainfall simulator for runoff plots. *Agric. Eng.* 39:644–648.

Meyer, L. D., and K. G. Renard. 1991. How research improves land management. In: *Agriculture and the Environment*. USDA Yearbook of Agriculture, pp. 20–27. U.S. Government Printing Office, Washington, DC.

Meyer, L. D., C. B. Johnson, and G. R. Foster. 1972. Stone and woodchip mulches for erosion control on construction sites. *J. Soil Water Conserv.* 27(6):264–269.

Meyer, L. D., G. R. Foster, and S. Nikolov. 1975a. Effect of flow rate and canopy on rill erosion. *Trans. ASAE* 18(5):905–911.

Meyer, L. D., G. R. Foster, and M. J. M. Römkens. 1975b. Sources of soil eroded by water from upland slopes. In: *Present and Prospective Technology for Predicting Sediment Yields and Sources*, pp. 177–189. USDA Agricultural Research Service ARS-S-40. USDA Science and Education Administration, Washington, DC.

Meyer, L. D., W. C. Harmon, and L. L. McDowell. 1980. Sediment sizes eroded from crop row sideslopes. *Trans. ASAE* 23:891–898.

Moldenhauer, W. C., and W. H. Wischmeier. 1960. Soil and water losses and infiltration rates on Ida Silt Loam as influenced by cropping systems, tillage practices, and rainfall characteristics. *Soil Sci. Soc. Am. Proc.* 24:409–413.

Monke, E. J., H. J. Marelli, L. D. Meyer, and J. F. DeJong. 1977. Runoff, erosion, and nutrient movement from interrill areas. *Trans. ASAE* 20:58–61.

Morgan, R. P. C. 1978. Field studies of splash erosion. *Earth Surf. Processes* 3:295–299.

Morgan, R. P. C. 1986. *Soil Erosion and Conservation*. Wiley, New York.

Morgan, R. P. C. 1991. Strategies for erosion control. In: D. A. Davidson (ed.), *Soil Erosion and Conservation*, pp. 162–185. Wiley, New York.

Moss, A. J., P. H. Walker, and J. Hutka. 1979. Raindrop-stimulated transportation in shallow water flows: an experimental study. *Sediment. Geol.* 22:165–184.

Mutchler, C. K. 1970. Size, travel, and composition of droplets formed by waterdrop splash on thin water layers. Ph.D. dissertation. University of Minnesota, St. Paul, MN.

Mutchler, C. K., and R. A. Young. 1975. Soil detachment by raindrops. In: *Present and Prospective Technology for Predicting Sediment Yields and Sources*, pp. 113–117. ARS-S-40. USDA Science and Education Administration, Washington, DC.

Mutchler, C. K., C. E. Murphree, and K. C. McGregor. 1994. Laboratory and field plots for erosion research. In: R. Lal (ed.), *Soil Erosion Research Methods*, 2nd ed., pp. 11–37. Soil and Water Conservation Society and St. Lucie Press, Ankeny, IA.

Napier, T. L. 1990. The evolution of U.S. soil conservation policy: from voluntary adoption to coercion. In: J. Boardman, J. D. L. Foster, and J. A. Dearing (eds.), *Soil Erosion on Agricultural Land*, pp. 627–644. Wiley, New York.

Napier, T. L., and S. M. Napier. 2000. Soil and water conservation policy within the United States. In: T. L. Napier, S. M. Napier, and J. Tvrdon (eds.), *Soil and Water Conservation Policies and Programs: Successes and Failures*, pp. 83–94. Soil and Water Conservation Society, Ankeny, IA.

Napier, T. L., S. M. Napier, and J. Tvrdon, (eds.) 2000, Soil and Water Conservation Society, Ankeny, IA.

National Research Council. 1999. *Hardrock Mining on Federal Lands*. National Academy Press, Washington, DC.

National Sedimentation Laboratory. 2001. Available on Web site: *www.sedlab.olemiss.edu/cwp_unit/acoustic.html*.

National Soil Erosion–Soil Productivity Research Planning Committee. 1981.

Soil erosion effects on soil productivity: a research perspective. *J. Soil Water Conserv.* 36(2):82–90.

Nearing, M. A., J. M. Bradford, and R. D. Holtz. 1987. Measurement of waterdrop impact pressures on soil surfaces. *Soil Sci. Soc. Am. J.* 51:1302–1306.

Nearing, M. A., G. Govers., and L. D. Norton. 1999. Variability in soil erosion data from replicated plots. *Soil Sci. Soc. Am. J.* 63:1829–1835.

Neff, E. L. 1979. Why rainfall simulation? In: *Proc. Rainfall Simulator Workshop*, Mar. 7–9, 1979, pp. 3–7. USDA Agricultural Research Service, Agricultural Reviews and Manuals, ARM-W-10/July 1979. U.S. Government Printing Office, Washington, DC.

Nikiforoff, C. C. 1942. Fundamental formula of soil formation. *Am. J. Sci.*, 240:847–866.

Nyhan, J. W., and F. Barnes. 1989. *Development of a Prototype Plan for the Effective Closure of a Waste Disposal Site in Los Alamos, New Mexico*. LA-11282-MS. Los Alamos National Laboratory, Los Alamos, NM.

Olson, K. R., R. Lal, and L. D. Norton. 1994. Evaluation of methods to study soil erosion–productivity relationships. *J. Soil Water Conserv.* 49(6):586–590.

Osterkamp, W. R., P. Heilman, and L. J. Lane. 1998. Economic considerations of a continental sediment-monitoring program. *Int. J. Sediment Res.* 13(4): 12–24.

Perrens, S. J., G. R. Foster, and D. B. Beasley. 1985. Erosion's effect of productivity along nonuniform slopes. In: *Erosion and Soil Productivity*, pp. 201–214. ASAE Publication 8–85. American Society of Agricultural Engineers, St. Joseph, MI.

Piest, R. F., J. M. Bradford, and R. G. Spomer. 1975. Mechanisms of erosion and sediment movement from gullies. In: *Present and Prospective Technology for Predicting Sediment Yields and Sources*. Publication ARS-S-40. pp. 162–176. USDA Science and Education Administration, Washington, DC.

Piest, R. F., R. G. Spomer, and P. R. Muhls. 1977. A profile of soil movement on a cornfield. In: *Soil Erosion: Prediction and Control*, pp. 160–166. Soil and Water Conservation Society, Ankeny, IA.

Pimental, D., F. J. Allen, A. Beers, L. Guinand, R. Linder, P. McLaughlin, B. Meer, D. Musonda, D. Perdue, S. Piosson, S. Siebert, K. Stoner, R. Salazar, and A. Hawkins. 1987. World agriculture and soil erosion. *Bioscience* 37: 277–283.

Pimental, D., C. Harvey, P. Resosudarmo, K. Sinclair, D. Kurz, M. McNair, S. Crist, L. Shpritz, L. Fitton, R. Saffouri, and R. Blair. 1995. Environmental and economic costs of soil erosion and conservation benefits. *Science* 267: 1117–1123.

Proffitt, A. P. R., P. B. Hairsine, and C. W. Rose. 1991. Rainfall detachment and deposition experiments with low slopes and significant water depths. *Soil Sci. Soc. Am. J.* 55:325–332.

Ree, W. O. 1949. Hydraulic characteristics of vegetation for vegetated waterways. *Agric. Eng.* 30:184–189.

Reeder, J. D., C. D. Franks, and D. G. Michunas. 2001. Root biomass and microbial processes. In: R. K. Follett, J. M. Kimble, and R. Lal (eds.), *The Potential of U.S. Grazing Lands to Sequester Carbon and Mitigate the Greenhouse Effect*. Lewis Publishers, Boca Raton, FL.

Renard, K. G. 1986. Rainfall simulators and USDA erosion research: history, perspective, and future. In: L. J. Lane, (ed.), *Proc. Rainfall Simulator Workshop*, Jan. 14–15, 1985, pp. 3–6. Society for Range Management, Denver, CO.

Renard, K. G., and G. R. Foster. 1983. Soil conservation: principles of erosion by water. In: *Dryland Agriculture*. Agronomy Monograph 23. *Am. Soc. Agron.* pp. 155–176.

Renard, K. G., G. R. Foster, G. A. Weesies, and J. P. Porter. 1991. RUSLE: revised universal soil loss equation. *J. Soil Water Conserv.* 46(1):30–33.

Renard, K. G., G. R. Foster, G. A. Weesies, D. K. McCool, and D. C. Yoder. 1997. *Predicting Soil Erosion by Water: A Guide to Conservation Planning with the Revised Universal Soil Loss Equation (RUSLE)*. USDA Agricultural Handbook 703. U.S. Government Printing Office, Washington, DC.

Renfro, G. W. 1975. Use of erosion equations and sediment-delivery ratios for predicting sediment yield. In: *Present and Prospective Technology for Predicting Sediment Yield and Sources*, pp. 33–45. USDA Agricultural Research Service, ARS-S-40. USDA Science and Education Administration, Washington, DC.

Ribaudo, M. O. 1986. *Reducing soil erosion: offsite benefits*. USDA Economic Research Service, Agricultural Economic Report 561. U.S. Government Printing Office, Washington, DC.

Richards, K. S. 1981. Introduction to morphometry. In: A. Goudie (ed.) *Geomorphological Techniques*. George Allen & Unwin, London.

Richardson, C. W. 1981. Stochastic simulation of daily precipitation, temperature, and solar radiation. *Water Resour. Res.* 17: 182–190.

Richardson, C. W., G. R. Foster, and D. A. Wright. 1983. Estimation of erosion index from daily rainfall amount. *Trans. ASAE* 26(1):153–156, 160.

Risse, L. M., M. A. Nearing, A. D. Nicks, and J. M. Laflen. 1993. Error assessment in the Universal Soil Loss Equation. *Soil Sci. Soc. Am. J.* 57: 825–833.

Ritchie, J. C., K. S. Humes, and M. A. Weltz. 1995. Laser altimeter measurements at Walnut Gulch Watershed, Arizona. *J. Soil Water Conserv.* 50(5): 440–442.

Roehl, J. W. 1965. *Sediment Source Areas, Delivery Ratios, and Influencing Morphological Factors*. Hydrology Publication 59, pp. 202–213. International Association of Hydrological Sciences, Wallingford, Berkshire, England.

Römkens, M. J. M., J. Y. Wang, and R. W. Darden. 1988. A laser microrelief meter. *Trans. ASAE* 31:408–413.

Salles, C., and J. Poesen. 1999. Performance of an optical spectro pluviometer in measuring basic rain erosivity characteristics. *J. Hydrol.* 218(3–4): 142–156.

Salvati, J. L., C. Huang, J. T. Johnson, and A. Klik. 2000. Soil microtopography from the Southern Great Plains Hydrology Experiments, 1999. In: *Conference Proc. International Geoscience and Remote Sensing Symposium* (IGARSS, 2000), pp. 1942–1944. IEEE Publications, Piscatoway, NJ.

Saxton, K., L. Stetler, and D. Chandler. 1997. Simultaneous wind erosion and PM_{10} fluxes. In: *Wind Erosion: An International Symposium / Workshop*, USDA, Wind Erosion Research Unit, June 3–5, Manhattan, KS.

Schafer, W. M., G. A. Nielsen, D. J. Dollhopf, and K. Temple. 1979. *Soil Genesis, Hydrological Properties, Root Characteristics and Microbial Activity of 1 to 50-Year-Old Stripmine Spoils*. EPA-600/7-79-100. U.S. Environmental Protection Agency, Cincinnati, OH.

Schertz, D. L., W. C. Moldenhauer, D. P. Franzmeier, and H. R. Sinclair, Jr. 1985. Field evaluation of the effect of soil erosion on crop productivity. In: *Erosion and Soil Productivity*, pp. 9–17. ASAE Publication 8–85. American Society of Agricultural Engineers, St. Joseph, MI.

Schmidt, B. L., R. R. Allmaras, J. V. Mannering, and R. I. Papendick (eds.). 1982. *Determinants of Soil Loss Tolerance*. American Society of Agronomy, Madison, WI.

Schumm, S. A. 1964. Seasonal variation of erosion rates and processes on hillslopes in western Colorado. *Z. Geomorphol.* Suppl. Bd. 5: 215–238.

Schumm, S. A. 1977. *The Fluvial System*. Wiley, New York.

Shainberg, I. 1992. Chemical and mineralogical components of crusting. In: M. D. Sumner and B. A. Stewart (eds.), *Soil Crusting:Chemical and Physical Processes*, pp. 33–53. Lewis Publishers, Boca Raton, FL.

Sharma, P. P., S. C. Gupta, and G. R. Foster. 1993. Predicting soil detachment by raindrops. *Soil Sci. Soc. Am. J.* 57:674–680.

Shaxson, T. F. 1981. Reconciling social and technical needs in conservation work on village farmlands. In: R. P. C. Morgan (ed.), *Soil Conservation: Problems and Prospects*, pp. 385–397. Wiley, New York.

Shields, F. D., Jr., A. J. Bowie, and C. M. Cooper. 1995. Control of streambank erosion due to bed degradation with vegetation and structure. *Am. Water Resour. Assoc. Water Resour. Bull.* 31(3): 475–489.

Shown, L. M., D. G. Frickel, R. F. Hadley, and R. F. Miller. 1981. *Methodology for Hydrologic Evaluation of a Potential Coal Mine: The Tsosie Swale Basin, San Juan County, New Mexico*. USGS Water Resources Investigation Open File Report 81–74. U.S. Government Printing Office, Washington, DC.

Simanton, J. R., R. M. Dixon, and I. McGowan. 1978. A microroughness meter for evaluating rainwater infiltration. *Hydrol. Water Resour. Ariz. Southwest* 8:171–174.

Simanton, J. R., E. Rawitz, and E. D. Shirley. 1984. Effects of rock fragments on erosion of semiarid rangeland soils. In: *Erosion and Productivity of Soils Containing Rock Fragments*, pp. 65–72. Soil Science Society of America, Madison, WI.

Simanton, J. R., M. A. Weltz, and H. D. Larsen. 1991. Rangeland experiments to parameterize the water erosion prodiction project model: vegetation canopy cover effects. *J. Range Manage*. 44(3):276–282.

Skaggs, R. W., L. F. Huggins, E. J. Monke, and G. R. Foster. 1969. Experimental evaluation of infiltration equations. *Trans. ASAE* 12(6): 822–838.

Skidmore, E. L. 1987. Wind-erosion climatic erosivity. *Clim. Change* 9: 195–208.

Skidmore, E. L. 1994. Wind erosion. In: R. Lal (ed.). *Soil Erosion Research Methods*, 2nd ed., pp. 265–293. Soil and Water Conservation Society and St. Lucie Press, Ankeny, IA.

Skidmore, E. L., and L. J. Hagen. 1977. Reducing wind erosion with barriers. *Trans. ASAE*, 20:911–915.

Skidmore, E. L., L. J. Hagen, D. V. Armburst, A. A. Dunar, D. W. Fryrear, K. N. Potter, L. E. Wagner, and T. M. Zobeck. 1994. Methods for investigating basic processes and conditions affecting wind erosion. In: R. Lal (ed.), *Soil Erosion Research Methods*, 2nd ed., pp. 295–330. Soil and Water Conservation Society and St. Lucie Press, Ankeny, IA.

St. Gerontidis, D. V., C. Kosmas, B. Detsis, M. Marathianou, T. Zafirious, and M. Tsara. 2001. The effect of moldboard plow on tillage erosion along a hillslope. *J. Soil Water Conserv.* 56: 147–152.

Staubitz, W. A., and J. R. Sobashinski. 1983. *Hydrology of Area 6, Eastern Coal Province, Maryland, West Virginia, and Pennsylvania*. USGS Water Resources Investigation Open File Report 83–33. U.S. Government Printing Office, Washington, DC.

Steinbeck, J. 1939. *The Grapes of Wrath*. Viking Press, New York.

Stout, J. E. 1990. Wind erosion in a simple field. *Trans. ASAE* 33:1597–1600.

Stout, J. E., and T. M. Zobeck. 1996. The Wolfforth field experiment: a wind erosion study. *Soil Sci. Soc. Am. J.* 16:616–632.

Sumner, M. D., and B. A. Stewart (eds.) 1992. *Soil Crusting: Chemical and Physical Processes*. Advances in Soil Science. Lewis Publishers, Boca Raton, FL.

Swanson, N. P. 1965. Rotating-boom rainfall simulator. *Trans. ASAE* 8:71–72.

Thomas, A. W., R. Welch, and T. R. Jordan. 1986. Quantifying concentrated-flow erosion on cropland with aerial photogrammetry. *J. Soil Water Conserv.* 41(4):249–252.

Toebes, C., and V. Ouryvaev (eds.). 1970. *Representative and Experimental Basins: An International Guide for Research and Practice*. UNESCO Henkes-Holland, Haarlem, The Netherlands.

Toy, T. J. 1983a. A comparison of the LEMI and erosion pin techniques. *Z. Geomorphol.* Suppl. Bd. 46:25–34.

Toy, T. J. 1983b. A linear erosion/elevation measuring instrument (LEMI). *Earth Surf. Processes Landforms* 8:313–322.

Toy, T. J., 1989. An assessment of surface-mine reclamation based upon sheet erosion rates at the Glenrock Coal Company, Glenrock, Wyoming. *Earth Surf. Processes Landforms* 14:289–302.

Toy, T. J., and J. P. Black. 2000. Topographic reconstruction: theory and practice. In: R. I. Barnhisel, R. G. Darmody, and W. L. Daniels (eds.), *Reclamation of Drastically Disturbed Lands* pp. 41–76. American Society of Agronomy, Crop Science Society of America, Soil Science Society of America, Madison, WI.

Toy, T. J., and W. L. Daniels. 1998. Reclamation of disturbed lands. In: R. A. Meyers (ed.), *Encyclopedia of Environmental Analysis and Remediation* pp. 4078–4100. Wiley, New York.

Toy, T. J., and R. F. Hadley. 1987. *Geomorphology and Reclamation of Disturbed Lands*. Academic Press, San Diego, CA.

Toy, T. J., and W. R. Osterkamp. 1999. The stability of rock-veneered hillslopes. *Intl. J. Sediment Res.* 14(3):63–73.

Trimble, S. W. 1977. The fallacy of stream equilibrium in contemporary denudation studies. *Am. J. Sci.* 277:876–887.

Trimble, S. W. 1983. A sediment budget for Coon Creek basin in the Driftless Area, Wisconsin, 1853–1977. *Am. J. Sci.* 283:454–474.

Trimble, S. W., and P. Crosson. 2000. U.S. soil erosion rates: myth and reality. *Science* 289:248–250.

Troeh, F. R., J. A. Hobbs, and R. L. Donahue. 1991. *Soil and Water Conservation* 2nd ed. Prentice Hall, Upper Saddle River, NJ.

Trout, T. J. 1996. Furrow irrigation erosion and sedimentation: on-field distribution. *Trans. ASAE* 39:1717–1723.

U.S. Agency for International Development (USAID). 2001. Available on Web site: *www.usaid.gov*.

U.S. Bureau of the Census. 2001. Available on Web site: *www.census.gov*.

U.S. Department of Agriculture, Forest Service. 1987. In: *Proc. Pinyon–*

Juniper Conference. General Technical Report INT-215. U.S. Government Printing Office, Washington, DC.

U.S. Department of Agriculture, Natural Resources Conservation Service. 1997a. *A Geography of Hope*. U.S. Government Printing Office, Washington, DC.

U.S. Department of Agriculture, Natural Resources Conservation Service. 1997b (revised 2000). *National Resources Inventory, 1997*, USDA Summary Report. Available on Web site: *www.nhq.nrcs.usda.gov/NRI/1997*.

U.S. Department of Agriculture, Natural Resources Conservation Service. 1997c. *Ponds: Planning, Design, Construction*. USDA Agricultural Handbook 590. U.S. Government Printing Office, Washington, DC.

U.S. Department of Agriculture, Natural Resources Conservation Service, 2001a, *Conservation Practice Standards*. Available on Web site: *www.nrcs.usda.gov/tech ref.*

U.S. Department of Agriculture, Natural Resources Conservation Service. 2001b. *National Handbook of Conservation Practices*. Available on Web site: *www.nrcs.usda.gov.*

U.S. Department of Agriculture, Natural Resources Conservation Service. 2001c. *National Soil Survey Handbook*. Available on Web site: *www.nrcs.usda.gov/tech ref.*

U.S. Department of Agriculture, Natural Resources Conservation Service. 2001d. *Soils*. Available on Web site: *www.nrcs.usda.gov.*

U.S. Department of Agriculture, Soil Conservation Service. 1972 (revised 1985). *National Engineering Handbook*, Sec. 4, *Hydrology*. U.S. Government Printing Office, Washington, DC.

U.S. Department of Agriculture, Soil Conservation Service. 1977. *Erosion and Sediment Control in Developing Areas: Planning Guidelines and Design Aids*. USDA Printing, Columbia, SC.

U.S. Department of Agriculture, Soil Conservation Service. 1978. *A Guide for Erosion and Sediment Control in Urbanizing Areas of Colorado*. U.S. Government Printing Office, Washington, DC.

U.S. Department of Agriculture, Soil Conservation Service. 1979. *Soil Survey of Santa Cruz and Parts of Cochise and Pima Counties, Arizona*. U.S. Government Printing Office, Washington, DC.

U.S. Department of Agriculture, Soil Conservation Service. 1983. *National Soils Handbook*, No. 430. U.S. Government Printing Office, Washington, DC.

U.S. Department of Agriculture, Soil Conservation Service. 1989 (revised 1994). *Soil Erosion by Wind*. Agricultural Information Bulletin 555. U.S. Government Printing Office, Washington, DC.

U.S. Environmental Protection Agency. 2000. *Storm Water Phase II Final*

Rule: Small Construction Program Overview. EPA-883-F-00–103. U.S. Environmental Protection Agency. Washington, DC. Also available on Web site: *www.epa.gov/NPDES.*

U.S. Geological Survey. 1997–1999. *National Field Manual for the Collection of Water-Quality Data*. USGS Techniques of Water-Resources Investigations, Book 9, Chaps. A1–A9, 2 vols. variously paged. U.S. Government Printing Office, Washington, DC.

Van Klaveren, R. W., and D. K. McCool. 1998. Erodibility and critical shear stress of a previously frozen soil. *Trans. ASAE* 41:1315–1321.

Van Liew, M. W., and K. E. Saxton. 1983. Slope steepness and incorporated residue effects on rill erosion. *Trans ASAE* 26:1738–1743.

Warrington, G. E., K. L. Knapp, G. O. Klock, G. R. Foster, and R. S. Beasley. 1980. Surface erosion. In: *An Approach to Water Resources Evaluation of Non-point Silvicultural Sources*, Chap. IV. EPA-160018-80-012. U.S. Environmental Protection Agency, Washington, DC.

Weber, T. A., and G. A. Margheim. 2000. *Conservation Policy in the United States: Is There a better Way?* In: Soil and Water Conservation Policies and Programs: Successes and Failures. T. L. Napier, S. M. Napier, and J. Tvrdon (eds.) Soil and Water Conservation Society, Ankeny, IA.

Weesies, G. A., S. J. Livingston, W. D. Hosteter, and D. L. Schertz. 1994. Effect of soil erosion on crop yield in Indiana: results of a 10 year study. *J. Soil Water Conserv*. 49(6):597–600.

Weltz, M. A., M. R. Kidwell, and H. D. Fox. 1998. Influence of abiotic and biotic factors in measuring and modeling soil erosion on rangelands: state of knowledge. *J. Range Manage*. 51:482–495.

West, N. W. 1990. Structure and function of microphytic soil crusts in wildland ecosystems of arid and semi-arid regions. *Adv. Ecol. Res*. 20:179–223.

Wind Erosion Research Unit, 2001. Available on Web site: *www.weru.ksu.edu/nres*.

Wischmeier, W. H. 1959. A rainfall erosion index for a Universal Soil-Loss Equation. *Soil Sci. Soc. Am. Proc*. 23:246–249.

Wischmeier, W. H. 1975. Estimating the soil loss equation's cover and management factor for undisturbed areas. In: *Present and Prospective Technology for Predicting Sediment Yields and Sources* pp. 118–124. USDA-ARS-S40. USDA Science and Education Administration, Washington, DC.

Wischmeier, W. H., and D. D. Smith. 1978. *Predicting Rainfall-Erosion Losses: A Guide to Conservation Planning*. USDA Agriculture Handbook 537. U.S. Government Printing Office, Washington, DC.

Wischmeier, W. H., C. B. Johnson, and B. V. Cross. 1971. A soil erodibility monograph for farmland construction sites. *J. Soil Water Conserv*. 26:189–193.

Wolman, M. G., and A. P. Schick. 1967. Effects of construction of fluvial sediment, urban and suburban areas of Maryland. *Water Resour. Res.* 3(2):451–464.

Woodruff, N. P. and F. H. Siddoway. 1965. A wind erosion equation. *Soil Sci. Soc. Am. Proc.* 29:602–608.

World Bank. 2001. *Environment*. Available on Web site: *www.worldbank.org/enviroment*.

Yalin, Y. S. 1963. An expression of bedload transportation. *J. Hydraul. Div. ASCE* 89:221–250.

Young, A. 1972. *Slopes*. Longman Group, London.

Young, R. A., and C. K. Mutchler. 1977. Erodibility of some Minnesota soils. *J. Soil Water Conserv.* 32:180–182.

Zachar, D. 1982. *Soil Erosion*. Elsevier Scientific Publishing, New York.

Zobeck, T. M. 1991. Abrasion of crusted soils: influence of abrader flux and soil properties. *Soil Sci. Soc. Am. J.* 55:1091–1097.

Zobeck, T. M. 2001, Field measurement of wind erosion. *Encyclopedia of Soil Science*. Marcel Dekker, New York.

Index